ROLE OF GLUTAREDOXIN AS A REDOX MODULATOR & ELUCIDATION OF CELL DEATH MECHANISM TRIGGERED BY MPTP

UZMA SAEED

ACKNOWLEDGEMENT

Memories of first day at NBRC are still afresh in my mind. Although it has been a long journey as a graduate student, nevertheless, it was pleasant, splendid and somewhat challenging. I am indebted to several people at NBRC who have been instrumental in molding my academic career. I would not have completed my doctorate successfully without their immense cooperation and support.

Undoubtedly the principal source of inspiration, my mentor and supervisor Professor Vijayalakshmi Ravindranath, to whom I express my deep and sincere gratitude for being extremely patient and supportive during this learning phase. Her dynamism, enthusiasm and rigor for research and her urge to elicit the best has left an everlasting impression on me. Without her, I would not have accomplished my goals. I owe her a lot for making me learn to constructively welcome and appreciate criticism, strive for the best and deliver an infallible performance. My salutations to her dedication for science and for contantly inspiring us to strive for better.

I truly thank Prof. Prasun Kumar Roy, Director NBRC for his cooperation and support. I also express my warm gratitude towards the members of my doctoral committee, Dr. Shyamala Mani, Dr. Anirban Basu and Dr. Shiv Kumar Sharma. Their critical evaluation and suggestions helped me a lot to perform better. I would also like to thank Dr. Nihar Ranjan Jana and Dr. Anirban Basu for helping me in my future prospects. My warm thanks to all the faculty, students and staff of National Brain Research Centre for helping me in some or the other way, and providing a harmonious and stimulating environment to learn and grow better.

I would also like to acknowledge the cooperation of my seniors Dr. Rajappa S Kenchappa, Dr. Latha Diwakar, Dr. Reddy P Kommaddi, Dr. Smitha Karunakaran, my dear labmates Varsha Agarwal, Alok Gupta,

Lalitha Durgadoss, Neha Sehgal, Ajit Ray for being a great team to work with. My special thanks to R. Khader Valli, Nidadavolu Prakash, Durga Praveen Meka, Shankar Datt Joshi, A. Madan Ram Kumar, Sujanitha Ramakrishnan and Nitin for their commendable help in my work. I am also extremely grateful to my friends and batchmates Shailesh Gupta, Shashank Tandon, Manisha Chugh and Amit Mishra for their encouragement and support during my Ph.D.

My special thanks to some people at NBRC who have influenced my life immensely and made my stay more enduring during the tough times. I owe my thanks to my closest friend and colleague, Nazia Khurshid, for always standing by me and encouraging me whenever failures made me lose hope. I also thank Nazia, Lalitha and Manoj Mishra for always being there whenever I needed a suggestion. I truly appreciate Neha Sehgal, for creating an extremely cordial aura whenever I found myself doomed by disappointments. I truly appreciate the caring spirit of Dr. Sudheendra Rao, whenever health was at stake. I salute them all for providing me with all the emotional support, camaraderie and care. This journey would not have been pleasant and easy without them.

I am extremely thankful to Dr. Shikha Yadav and staff of animal research facility, NBRC for their help and cooperation. I also wish to thank Hariharan Krishnan for the excellent support from stores and purchase section; PVS Shyam Kumar and K.V.S Kameswara Rao from academics department; Jothi Basu, Mahender Singh, Kedar Singh Bagecha and Ganesh Gurumoorthy from Distributed Information Centre (DIC). I would also like to thank Pooja Gosain and Debashish Bhattacharjee for their help with secretarial work and Mr. P. Manish for his help in the laboratory.

My affectionate regards towards my family. I deeply acknowledge the support and understanding of my parents and my brothers. I also appreciate the patience of my friend and cousin Shezy, to whom I poured, each time I was in distress.

I greatly acknowledge the financial support of National Brain Research Centre and University Grants Commission (Council for Scientific & Industrial Research) for supporting me financially throughout my doctorate. I am also thankful to Department of Biotechnology (DBT) for financially supporting my travel grant to attend an international conference.

Finally, my utmost thanks to GOD for giving me this life, the privilege to be at NBRC and for all the wonderful and caring people around me, who influenced my life remarkably. I thank you GOD for all your eternal blessings. May your blessings, always be bestowed on each of us.

Thank you

To my dear parents, my teachers and my Ph.D. mentor .....

..... For molding me into what I am.

TABLE OF CONTENTS

	PAGE NUMBER
ABBREVIATIONS	1 - 4
INTRODUCTION	5 - 68
MATERIALS AND METHODS	69 - 119
RESULTS	120 - 242
DISCUSSION	243 - 271
SUMMARY AND CONCLUSION	272 – 278
BIBLIOGRAPHY	279 - 323

INTRODUCTION

- **1.1 Neurodegenerative disorders**
  - 1.2 Neurodegeneration in AD
  - 1.3 Neurodegeneration in HD
  - 1.4 Neurodegeneration in ALS

- **2.0 Parkinson's disease**
  - 2.1 Epidemiology
  - 2.2 Etiology
  - 2.3 Pathogenesis in PD
  - 2.4 Genes contributing to pathogenesis of PD
  - 2.5 Animal models used in PD

- **3.0 Cell death in PD**
  - 3.1 Cell death pathways underlying pathogenesis of PD
  - 3.2 Mechanisms of neurodegeneration in PD
  - 3.3 Protein aggregation and Ubiquitin Proteasome System
  - 3.4 Mitochondrial dysfunction and oxidative stress
  - 3.5 Modes of cell death

- **4.0 Role of mitochondria in PD**
  - 4.1 Mitochondrial inheritance and genetics in PD
  - 4.2 Oxidative stress and Mitochondria
  - 4.3 Targets of ROS (DNA, protein, lipids)
  - 4.4 Dopamine as an inducer of reactive oxygen species
  - 4.5 Kreb cycle dysfunction and inactivation of Fe-sulfur proteins
  - 4.6 Effect of ROS on Ca^{2+} dependent mitochondrial permeability transition

- **5.0 Mitochondrial membrane permeability transition**
  - 5.1 Maintenance of MMP by electron transport chain
  - 5.2 Permeabilization of outer mitochondrial membrane
  - 5.3 Mechanism of permeabilization
  - 5.4 Excitotoxicity
  - 5.5 Afferent signals from other organelles affecting MPT and cell death effectors
  - 5.6 Mitochondrial cell death effectors
  - 5.7 Mitochondrial permeability transition and neurodegeneration

- **6.0 Regulation of mitochondrial ROS production**
  - 6.1 Glutathione
  - 6.2 Thiol disulfide oxidoreductases
  - 6.3 Glutaredoxin system
  - 6.4 Role of glutaredoxin
  - 6.5 Thioredoxin (Trx) system
  - 6.6 Role of redox perturbation in mitochondrial dysfunction
  - 6.7 Redox-mediated signaling

- **6.8** ASK1-JNK-Daxx pathway
- **6.9** ASK1-p38 pathway
- **6.10** Redox modulating enzymes and regulation by estrogen

7.0 Estrogen mediated neuroprotection in PD
- **7.1** Epidemiology
- **7.2** Estrogen as a neuroprotective molecule
- **7.3** Estrogen receptors (ERs), structure and function
- **7.4** Mode of neuroprotective action
- **7.5** Role of estrogen in PD
- **7.6** Hormone replacement therapy

8.0 Treatment strategies for PD and milestones covered so far
- **8.1** Reasons for failures and future perspectives

1.0 Neurodegenerative disorders: Neurodegenerative diseases constitute a vast group of chronic, progressive and heterogeneous brain disorders which are characterized by selective degeneration of subsets of neurons, or their dysfunction in specific areas of central nervous system (CNS). This gradual and progressive degeneration leads to either motor impairment or cognitive decline depending on the type of neuronal population affected. These neurodegenerative disorders include complex multifactorial pathologies such as Alzhiemer's disease (AD), Parkinson's disease (PD) and Amyotrophic Lateral Sclerosis (ALS) and monogenic disorders such as Huntington's disease (HD). Symmetric loss of neurons observed in these disorders may be due to the gene-environment interplay, although the specific causes underlying each are not clear. Both, predisposition of susceptibility genes and environmental factors are responsible and a fine interplay among these two appear to lead to the vulnerability to the disease (Migliore and Coppede, 2002).

Besides fully penetrant causative genes responsible for inherited forms of neurodegenerative disorders, several susceptibility genes have been identified as risk factors for sporadic forms of these diseases. Moreover, environmental factors including pesticides, metals, head injury and infection have also been studied as potential risk factors for neurodegeneration and an interaction between environmental factors and genetic polymorphisms of metabolic enzymes has been observed (Brown et al., 2005). Oxidative damage and mitochondrial impairment have been extensively observed in neurodegenerative pathologies, since oxidative stress and mitochondrial dysfunction are the hallmark for most of the neurodegenerative disorders (Reichmann et al., 1995). Mitochondria themselves may often act as key producers of reactive oxygen species (ROS) and consequently lead to oxidative stress thereafter, thus causing neurodegeneration during ageing, which is also a risk factor for neurodegeneration. However, the genetic basis of some of these neurodegenerative disorders has been elucidated, the causative mechanisms of sporadic form of the disease, which is more prevalent, is poorly understood.

Although all neurodegenerative conditions are characterized by oxidative stress, the neuronal population affected in each case is specific to the disorder. The death of dopaminergic neurons in substantia nigra pars compacta (SNpc) underlies Parkinson's disease (Allain and Bentue-Ferrer, 1997). Betz cells in motor cortex and anterior horn cells in the spinal cord are lost in amyotrophic lateral sclerosis (Price et al., 1987; Sriram et al., 1998). Atrophy of hippocampal and cortical neurons is observed in Alzheimer's disease (Scheibel, 1979; Goedert, 1987) whereas Huntington's disease results from loss of neurons in the striatum (Jakel and Maragos, 2000). The major challenge is to understand why, specific subset of neurons degenerate in specific brain regions in different neurodegenerative disorders. The loss of specific neuronal populations in these neurodegenerative disorders is thought to occur through necrosis and programmed cell death (PCD; (Offen et al., 2000; Elmore, 2007; Lorz and Mehmet, 2009)). PCD in neurodegenerative conditions can be triggered by a variety of cellular insults such as misfolded proteins, reactive oxygen and nitrogen species, inhibition of mitochondrial complex I, calcium influx, excitotoxicity, trophic-factor withdrawal, and death-receptor activation. Although the role of PCD has been controversial, studies targeting PCD by inhibiting caspases has been successful thereby indicating its role in neurodegeneration (Wales et al., 2008). Enormous work has been done towards understanding the causative mechanism underlying demise of specific subsets of neurons in neurodegenerative conditions but things are not yet clear. The current work is an attempt to understand the role of oxidative stress, mitochondrial dysfunction and cell death signaling in neurodegeneration involved in PD since understanding the underlying molecular pathogenesis will help in designing rationale based effective therapies.

1.1 Neurodegeneration in Alzhiemer's disease: Alzhiemer's disease (AD) is the most common and devastating neurodegenerative illness which accounts for majority of age related dementia. It is characterized by global cognitive decline and accumulation of amyloid-β (Aβ) deposits and neurofibrillary tangles in the brain which are the neuropathological hallmarks for the disease (1987). Genetically, 160 mutations in 3 genes have been reported to cause early onset familial AD. These include Aβ precursor protein (APP;(Goate et al., 1991)), Presenilin 1 (PSEN1;(Sherrington et al., 1995)) and Presenilin 2 (PSEN2; (Levy-Lahad et al., 1995; Rogaev et al., 1995)). Late onset AD represents the majority of AD cases wherein ε-4 allele of apolipoprotein gene (APOE-ε4) acts as a genetic risk factor. While the AD linked genes are located on different chromosomes, they all seem to share a common biochemical pathway, i.e. altered production of amyloid-β-42 (Aβ42) leading to its relative overabundance over Aβ40, thereby subsequently resulting in neuronal cell death in hippocampus, entorhinal cortex and association areas of neocortex, thus causing dementia. Oxidative damage and mitochondrial dysfunction have been widely implicated in pathogenesis of AD. A consistent impairment of mitochondrial enzyme, i.e. a defect in electron transport chain (ETC) carrier cytochrome c oxidase (COX; complex IV), has also reported in AD (Manczak et al., 2006; Pope et al., 2008). This results in limited ATP and enhanced reactive oxygen species (ROS) generation causing mitochondrial dysfunction thereby triggering apoptosis (Beal, 1998; Ding et al., 2008).

1.2 Neurodegeneration in Huntington's disease: Huntington's disease (HD) is a monogenic autosomal disorder characterized clinically by chorea, psychiatric disturbances and dementia (Davies and Ramsden, 2001). Physiologically, it is characterized by selective degeneration of medium spiny GABAergic neurons in striatum. HD is caused due to expansion of CAG repeats (encoding glutamine – poly-Q expansion) within exon 1 of Huntingtin gene (Gusella et al., 1983). The expanded poly-Q segment confers a dominant 'gain of function' leading to protein misfolding and formation of inclusion bodies and neurodegeneration. Involvement of mitochondrial dysfunction and impairment of energy production is suggested in HD (Gardian and Vecsei, 2004). Oligomerization of expanded polyglutamine triggers cell death mechanisms such as release of cytochrome c from the mitochondria and activation of pro-apoptotic initiators such as caspase 1 and 8, which have been reported in HD.

1.3 Neurodegeneration in Amyotrophic Lateral Sclerosis & other motor neuron diseases: Amyotrophic Lateral Sclerosis (ALS) is the most common motor neuron disorder characterized by selective degeneration of anterior horn cells of spinal cord and cortical motor neurons leading to progressive neuromuscular dysfunction, paralysis and subsequent death of the patient. It is predominantly sporadic but 10% of the cases are familial and being attributed mainly to mutations in genes encoding for Cu/Zn cytosolic superoxide dismutase 1 (SOD1/ALS1; (Rosen, 1993; Rosen et al., 1993)), Alsin (ALS2) and TDP-43 (Gros-Louis et al., 2006; Sreedharan et al., 2008). Mutations in SOD1 gene is suggested to cause neurodegeneration by impaired oxidative stress response, protein misfolding, cytoskeletal dysfunction, perturbed Ca^{2+} ion homeostasis involving glutamate related excitotoxic component (Stathopulos et al., 2003; Wood et al., 2003) and triggering of apoptotic signaling cascade. Studies with transgenic animal models of ALS harboring mutations in SOD1 indicate that motor neuron degeneration is caused by toxic gain of function of the mutated SOD protein.

2.0 Parkinson's disease: Parkinson's disease (PD), first reported by James Parkinson in 1817, is the most common movement disorder known. It is also the second most common age related neurodegenerative disorder after Alzheimer's disease, affecting 1% of population above 65 yr (Lang and Lozano, 1998b). Clinically, PD is represented by cardinal symptoms of bradykinesia, resting tremor, rigidity and postural instability and is also accompanied by autonomic, cognitive and psychiatric disturbances as the disease progresses. Pathological features include degeneration of dopaminergic neurons in the substantia nigra pars compacta (SNpc) of the ventral midbrain, and loss of their terminals in the striatum which is more pronounced. At the time of diagnosis, approximately 50-70% of dopaminergic neurons are lost (Hornykiewicz, 1975) and the existing neurons in PD brain contain (Lewy bodies) LBs, which are intracytoplasmic inclusions containing neurofilaments, ubiquitin and α-synuclein among others (Spillantini et al., 1997). Consequent to the information that loss of SNpc neurons causes dopamine (DA) deficiency in the striatal terminals and replenishing DA could alleviate most of the symptoms in PD, discovery of levodopa revolutionized the treatment, until several years later, when patients treated with the drug developed 'dyskinesias'. Current research is directed towards slowing down the progressive degeneration of dopaminergic neurons 'neuroprotective' and/or restoring sick neurons to their normal state by 'neurorestorative' therapy.

2.1 Epidemiology: The prevalence of PD in India and USA is about 260 and 550 people per 100,000 respectively over the age of 65 years (Dorsey et al., 2007). It occurs throughout the world, in all ethnic groups, affects both sexes (Zhang and Roman, 1993), although the relative risk being 1.5 times greater in men than women with an overall male:female (M:F) ratio of 1.46 (95% confidence intervals 1.24, 1.72; (Burn, 2007)). Approximately 5 to 10% of patients develop symptoms before the age of 40 (early-onset PD; (Quinn et al., 1987)). The lowest reported incidence is among Asians and African americans, whereas the highest is among American whites (Jendroska et al., 1994). Thus, this epidemiological pattern suggests that the propensity for the development of

Parkinson's disease is universal but that local environmental factors may have a role in augmenting the vulnerability.

2.2 Etiology: The etiology of PD is ambiguous with uncertainty about the relative contribution of environmental and genetic factors; however, animal models, human postmortem tissue analysis and genetic analyses have provided important clues. Some familial forms of autosomal-dominant PD are due to mutations (A53T and A30P) in α-synuclein (Conway et al., 1998; Narhi et al., 1999). Mutations in the ubiquitin E3 ligase, parkin, cause autosomal-recessive PD (Fishman and Oyler, 2002), and mutations in ubiquitin C-terminal hydrolase L1 (UCH-L1) may cause autosomal-dominant PD (Gasser, 2001). A mutation in the medium neurofilament subunit in a patient with early-onset severe PD suggests that aberrations in the cytoskeleton may be linked to PD. These rare familial cases of PD have shed light on the more common sporadic form. According to environmental hypothesis, neurodegeneration in PD results from dopaminergic neurotoxins which have been shown to cause similar selective neuronal loss and parkinsonism, such as 6-hydroxy dopamine, rotenone, paraquat and most importantly, 1-methyl-4-phenyl-1,2,3,6-tetrahydropyridine (MPTP). Both familial and sporadic forms present with similar clinical and pathological symptoms in PD. This supports the hypothesis that genetic susceptibility factor(s) may exist which are influenced by environmental factors and which thus increase the risk of PD; thus potraying that PD, probably is caused by more than one etiological factor.

2.3 Pathogenesis in PD: Onset of PD may be initiated by either genetic or environmental factors, which may lead to Parkinsonian symptoms. Whatever the cause may be, studies with genetic and toxin animal models have suggested two major hypotheses propagating the pathogenesis of the disease – one of them proposing misfolding and aggregation of protein which is instrumental in the death of SNpc dopaminergic neurons while according to the other hypothesis mitochondrial dysfunction and consequent oxidative stress is the key culprit. Both these pathogenic pathways work concurrently and the current PD research is directed to elucidate the sequence in which they act and whether their interaction at certain points is the key to the demise of SNpc dopaminergic neurons. One such evidence is the finding that α-synuclein after oxidative damage becomes more prone to protein misfolding and aggregation (Giasson et al., 2000). There is ambiguity whether multiple cell death pathways that are stimulated during neurodegeneration in PD finally either converge to similar downstream cell death machinery such as apoptosis or remain divergent. Further probing into the mode and downstream mechanism of cell death may be of great consequence in proposing therapeutic targets for PD.

2.4 Genes contributing to pathogenesis of PD: Several genes have been identified and linked to familial monogenic PD and all these genes encode proteins of seemingly diverse function. Among them are E3 ubiquitin ligase (Parkin), mitochondrial serine/threonine kinase (PINK1), redox regulated chaperone (DJ-1), presynaptic protein (α-synuclein) and leucine-rich repeat kinase 2 (LRKK2). Several of these genes function within mitochondria and help in reducing oxidative stress thus suggesting that both genetic mutations as well as post-translational modification of proteins implicated in PD are linked to the pathogenesis, hence the pathogenic mechanism underlying both familial and sporadic PD are similar.

i. **Parkin:** Mutations in Parkin gene cause autosomal recessive juvenile-onset Parkinsonism (AR-JP; (Kitada et al., 1998; Abbas et al., 1999)). Parkin functions as E3-ubiquitin ligase in proteasome mediated degradation of several proteins which accumulate in brain of AR-JP patients. It functions as a potent mitochondrial protection factor and its overexpression helps in maintenance of mitochondrial membrane potential, enhances the expression of subunits of mitochondrial complex I selectively and reduces the accumulation of ROS within mitochondria (Kuroda et al., 2006a). Drosophila knockout for parkin are more prone to oxidative insults and develop mitochondrial pathology (Greene et al., 2003; Pesah et al., 2004). Also, parkin deficient mice show increased protein oxidation, lipid peroxidation and decreased expression of subunits of mitochondrial complex I and IV, thereby, causing mitochondrial dysfunction. In proliferating cells parkin is known to enhance transcription of mtDNA thus stimulating mitochondrial biogenesis during mitosis (Kuroda et al., 2006b). It is also reported to be recruited to damaged dysfunctional mitochondria where it plays a crucial role in mitochondrial turnover by mitophagy (McBride, 2008; Narendra et al., 2008).

ii. **α-synuclein:** It was the first gene identified with three missense mutations (A53T, A30P, E46K) and a triplication of the *α-synuclein* gene locus (*SNCA, PARK1*) leading to dominantly inherited early-onset PD (Polymeropoulos et al., 1997; Kruger et al., 1998; Bradbury, 2003; Singleton et al., 2003; Zarranz et al., 2004). It is a component of LBs, is ubiquitously expressed in mammalian brain, enriched in presynaptic terminals where it remains associated with membrane and vesicular structures (Irizarry et al., 1996; Kahle et al., 2000). Although, it's physiological function is not clear but it is known to be involved in synaptic vesicle recycling and DA neurotransmission (Abeliovich et al., 2000). Mutations and its overexpression accelerate the formation of protofibrils (Conway et al., 1998; Conway et al., 2000; Fredenburg et al., 2007), which are stabilized by dopamine (Xu et al., 2002) thus causing toxicity due to oxidative stress, proteasome inhibition, endoplasmic reticulum stress and mitochondrial dysfunction. Overexpression of wild type or mutants also lead to mitochondrial abnormalities such as oxidation of mitochondrial protein (Poon et al., 2005), mtDNA damage, reduced complex IV activity (Martin et al., 2006), increased sensitivity to MPTP (Song et al., 2004) and release of cytochrome c, activation of caspases and subsequent cell death. Association of α-synuclein with complex I in inner mitochondrial membrane and its high levels in mitochondria from substantia nigra (SN) of PD patients indicate that its accumulation in mitochondria may cause complex I inhibition and oxidative stress (Devi et al., 2008).

iii. **UCH-L1:** UCH-L1 encodes ubiquitin carboxyl terminal hydrolase L1 which harbour heterozygous mutations I93M and M124L, identified in affected individuals. It is a highly abundant neuron specific protein, which is categorized as deubiquitinating enzyme responsible for hydrolyzing polymeric ubiquitin chains to free ubiquitin monomers (Wilkinson et al., 1989) and also promote their stability in vivo (Osaka et al., 2003). It also functions as ubiquitin-proteasome ligase (Liu et al., 2002) and has been localized to LBs in sporadic PD (Lowe et al., 1990). Significance of its mutations in PD is not well understood, however, I93M mutation leads to reduced *in vitro* hydrolytic activity of UCH-L1, thus causing a loss of function. It is also known to cause the accumulation of α-synuclein in cultured cells through the addition of Lys63 linked polyubiquitin chain (Liu et al., 2002). Hence both ligase and hydrolase activity of UCH-L1 helps in maintaining the normal function of UPS and a loss of function may be implicated in the pathogenesis of PD.

iv. **Leucine-rich repeat kinase 2 (LRRK2):** LRRK2 mutations are responsible for majority of familial PD including autosomal-dominant PD and are linked to sporadic late-onset PD (Paisan-Ruiz et al., 2004; Zimprich et al., 2004). LRRK2 is a serine/threonine kinase with conserved mitogen-activated protein kinase kinase kinase (MAPKKK) domain. Over-expression of mutant LRRK2 induces caspase dependent neuronal death (Iaccarino et al., 2007) and mutations augment the kinase activity, which further mediates neurotoxicity (Gloeckner et al., 2006; Greggio et al., 2006; Smith et al., 2006). Overexpression of G2019S mutant of LRRK2 in neuroblastoma cells causes retraction and shortening of neurites and inhances autophagy (Plowey et al., 2008). LRRK2 has been normally localized within mitochondria and lysosomes in mammalian brain, and is also associated with intracellular membranes, including transport vesicles (Biskup et al., 2006). When overexpressed, 10% of wild-type and mutant LRRK2 is found in mitochondrial fraction thereby indicating its capability to induce apoptosis and compromise mitochondrial function.

v. **PTEN-induced putative Kinase I (PINK1):** Mutations in PINK1 (*PARK6*) are associated with hereditary early-onset PD, heterozygous mutations are implicated in sporadic early-onset PD (Valente et al., 2004a) and cause loss of function. It is a highly conserved serine/threonine kinase of Ca^{2+} calmodulin family of protein, with a mitochondrial targeting sequence at its N-terminus (Valente et al., 2004b; Silvestri et al., 2005). PINK1 affords protection against mitochondrial dysfunction and apoptosis induced by proteasomal inhibition. Its protective function is impaired by G309D mutation localized in its ADP binding site, important for ADP binding and kinase activity (Bossy-Wetzel et al., 2004; Valente et al., 2004b). TNF-receptor-associated protein 1 (TRAP1) and HtrA2/Omi, both present in the mitochondrial intermembrane space have been identified its substrate (Plun-Favreau et al., 2007; Pridgeon et al., 2007). PINK1 phosphorylates TRAP1, a chaperone which in turn prevents oxidative stress-induced release of cytochrome *c* and apoptosis (Pridgeon et al., 2007). On activation of p38 MAPK pathway, PINK1 phosphorylates HtrA2/Omi thus increasing its protease activity and mediating survival under stress (Plun-Favreau et al., 2007). In human brains with sporadic PD, phosphorylation of HtrA2 at serine 142 is significantly increased whereas it is missing in brains of PD patients with PINK1 mutations (Plun-Favreau et al., 2007). It has also been reported recently that over-expression of PINK1 in the cytosol or deletion of its mitochondrial targeting sequence also efficiently protects neurons against MPTP suggesting that PINK1 may also have cytosolic substrates which have an impact on mitochondrial function (Haque et al., 2008).

vi. **DJ-1:** DJ-1 (PARK7) gene encodes a highly conserved 189 amino acid protein belonging to DJ-1/Thij family; mutations in it cause recessively inherited PD (Bonifati et al., 2003; Hague et al., 2003). Several homozygous, heterozygous and missense pathogenic mutations, identified so far are L166P, M26I and D149A. DJ-1 is ubiquitously expressed in all mammalian tissues, in brain it is localized both in neuron and glia (Bandopadhyay et al., 2004; Olzmann et al., 2004). It is a multifunctional protein which acts as an antioxidant, chaperone and transcriptional modulator. Its knockdown in neurons triggers cell death induced by oxidative stress, ER stress and proteasome inhibition (Yokota et al., 2003; Taira et al., 2004). DJ-1 is physiologically present mostly in the cytosol and nucleus but under conditions of oxidative stress, it gets translocated to mitochondria and confers neuroprotection (Canet-Aviles et al., 2004; Ashley et al., 2009; Junn et al., 2009). Oxidative stimuli such as MPP^+, paraquat, rotenone and 6-hydroxydopamine trigger redistribution of DJ-1 in mitochondria, following oxidation of thiol group present in cysteine residue at position 106 to cysteine-sulfinic acid (Canet-Aviles et al., 2004; Blackinton et al., 2009). Several studies suggest that DJ-1 scavenges reactive oxygen species (ROS) and by doing so it helps maintain the redox status of other proteins (Mitsumoto et al., 2001; Ooe et al., 2005; Lev et al., 2008). DJ-1 has multiple cysteine residues whose oxidation enables it to act as a redox sensor during oxidative stimuli, such as hydrogen peroxide or paraquat (Mitsumoto et al., 2001; Taira et al., 2004). Synthetic mutations of the cysteines at positions 46, 53 and 106 have provided an understanding of the response of DJ-1 to oxidative stress leading to cellular protection. It adds increasing number of oxygen atoms to its cysteine sulfydryl groups leading to alteration in its localization and function. DJ-1, upon oxidation of its C106 residue undergoes cleavage between glycine and proline at amino acid 157 and 158 (Ooe et al., 2006). Studies with synthetic mutations of DJ-1 suggest C106 to be the most vulnerable site for oxidation when compared to C46 and C53. C106 is reversibly modified to sulfenic acid (R-S-OH) when exposed to oxidative stimuli such as hydrogen peroxide, rotenone or MPP$^+$ (Canet-Aviles et al., 2004; Kinumi et al., 2004). The C106A mutation also restricts its translocation to mitochondria and subsequent protection against oxidative stress (Canet-Aviles et al., 2004). It also acts as a redox sensitive molecular chaperone which is activated during oxidative conditions and α-synuclein being one of its substrate (Shendelman et al., 2004). MPTP induced degeneration of dopaminergic neurons is further enhanced in DJ-1-knockout mice (Kim et al., 2005b), whereas its over-expression via viral gene delivery in the nigrostriatal system attenuates MPTP-induced cell death of dopaminergic neurons (Kim et al., 2005b; Paterna et al., 2007). DJ-1 interacts with PINK1 and increases its steady-state levels (Tang et al., 2006), as well as with Parkin under oxidative conditions (Moore et al., 2005). It also increases cellular GSH levels through an increase in glutamate cysteine lygase, a rate limiting enzyme for GSH synthesis. It also acts as a transcriptional co-activator and interacts with nuclear proteins in dopaminergic neuronal cells and protects against neuronal apoptosis (Xu et al., 2005). DJ-1 is also implicated in apoptosis since it interacts with Daxx (a death associated protein), and sequesters it within the nucleus thus preventing it from gaining access to cytoplasm, associating with and thereby activating its effector kinase apoptosis signal regulating kinase 1 (ASK1) and triggering the ensuing cell death cascade (Junn et al., 2005).

vii. **HtrA2/Omi:** HtrA2/Omi has been previously implicated in neurodegeneration and is recently being associated with predisposition to PD since a heterozygous G399S mutation and an A141S polymorphism have been reported (Strauss et al., 2005). It is a serine protease harboring PDZ domain and an N-terminal targeting motif. It is localized in mitochondrial intermembrane space and is released into the cytosol during apoptosis where it binds to inhibitor of apoptotic proteins (IAPs) thus relieving the caspases free (Suzuki et al., 2001; Seong et al., 2004). It also induces cell death through its proteolytic activity. HtrA2 reportedly has a neuroprotective role as supported by *in vivo* studies and loss of HtrA2/Omi protease activity by the Ser276Cys mutation results in neurodegeneration and also increases the propensity of permeability transition in mitochondria and stress-induced cell death (Jones et al., 2003). HtrA2 knockout mice are also known to develop striatal neurodegeneration with parkinsonian phenotype (Martins et al., 2004). Since HtrA2/Omi is a substrate for PINK1, genetic interaction among PINK1 and HtrA2/Omi has been suggested (Whitworth et al., 2008). However, unlike deficiency of PINK1 and Parkin, the loss of HtrA2/Omi in Drosophila did not cause mitochondrial abnormalities, apoptotic muscle degeneration and dopamine neuron loss (Yun et al., 2008).

While there has been substantial progress over the past decade in identification of PD linked genes and in our understanding about their genetics in pathogenesis using transgenic and knockout models, the etiological factors of the disease are still not clear. Genetic information has essentially helped to elucidate the molecular and biochemical pathways and has also focused on the emerging role of oxidative stress and mitochondrial dysfunction as early events in the neurodegenerative process. Further studies investigating the role of mitochondrial pathology in neurodegeneration will help to understand the interplay between environmental and genetic factors and their respective roles in neurodegeneration.

2.5 Animal models used in PD: Animal models offer an invaluable tool for understanding disease pathogenesis. A relevant animal model for PD should possess the following attributes:-

i. **Behavioral and clinical correlation:** It should be a behavioral correlate of the nigrostriatal dopaminergic pathway degeneration and should possess clinical manifestations of the disease such as rigidity, tremor, bradykinesia and postural instability, however, only non human primates but not rodent models show such behaviour.

ii. **Selective vulnerability of dopaminergic neurons:** Dopaminergic neurons residing in substantia nigra pars compacta (SNpc; A9 subgroup) and their terminals in striatum degenerate whereas dopaminergic neurons of ventral tegmental area (VTA) are unaffected in PD (German et al., 1988), such selective pathology should be mimicked in an animal model for PD.

iii. **Slow, progressive and age-dependent loss of dopaminergic neurons:** PD involves slow and progressive degeneration of dopaminergic neurons which is linked to ageing. Animal model for PD should show a presymptomatic phase with a moderate lesion, followed by a symptomatic phase where the lesions reach a threshold ultimately leading to dopaminergic denervation with more pronounced clinical symptoms.

iv. **Progressive, age-dependent degeneration of non dopaminergic neurons:** Animal model should manifest non-dopaminergic lesions also in the brain at later stages of pathology which is accompanied by the manifestations of non- motor symptoms.

v. **Complex I deficiency:** PD pathogenesis is also characterized by complex I deficiency and mitochondrial dysfunction, a desirable trait in an animal model.

vi. **Lewy bodies (LBs) like inclusions:** LBs are hallmark of PD therefore an animal model should preferably possess intracytoplasmic proteinaceous inclusions both in dopaminergic and non-dopaminergic neurons showing immunoreactivity for α-synuclein and ubiquitin.

vii. **Responsiveness to L-DOPA:** L-DOPA helps in relieving the symptoms, hence the animal model should respond to L-DOPA.

Genetic models of PD: Although majority of PD cases are sporadic, a fraction of them are familial too. The discovery of genes linked to familial forms of PD has helped in development of novel genetic mouse or Drosophila models overexpressing wild type or mutants of α-synuclein (Masliah et al., 2000; van der Putten et al., 2000; Lee et al., 2001; Giasson et al., 2002), Parkin knockout (Itier et al., 2003; Palacino et al., 2004), DJ-1 knockout (Goldberg et al., 2005; Kim et al., 2005b), Nurr1 heterozygote (Jiang et al., 2005), and Pitx3 aphakia (Hwang et al., 2005). Numerous mouse models created for α-synuclein, with either overexpressed human wild-type gene, reproducing the gene duplication and triplication normally detected in PD families, or the PD-causing α-synuclein mutations A30P or A53T expressed in transgenic mice (Richfield et al., 2002), have imparted valuable information to understand the pathogenic mechanisms involved in familial PD. High levels expression of mutated α-synuclein under the mouse prion protein promoter induced a progressive phenotype with intraneuronal inclusions, degeneration and mitochondrial DNA damage in the neurons (Martin et al., 2006). Even though the key symptoms for PD were not reproduced in this model, it is valuable for understanding the relationship of α-synuclein positive protein depositions and neuronal damage. Inspite of all the effort put in, none of the genetic models generated so far, fully recapitulate the significant pathological features of the disease such as loss of dopaminergic neurons. However, certain subtle effects on dopamine system have been detected such as slight reduction in the levels of striatal dopamine, decrease in dopamine transporter (DAT) binding and some motor deficits. Enhanced substantia nigra vulnerability towards MPTP has been observed in several genetic models but these models show variability in motor deficits and response to L-DOPA.

Tissue specific knockout models: The increasing trend of knockout animal models with disrupted genes in a specific region of interest or in a selective population of neurons have led to the development of dopamine neuron specific knockouts like DAT-cre, Ret lox/lox, the mitochondrial transcription factor A (TFAM) and Dicer lox/lox, that are promising preclinical models with slow progression (Terzioglu and Galter, 2008). Although, these models have not been potentially exploited for research due to their complex breeding scheme, they may be excellent to study post-transcriptional mechanisms in pathogenesis of PD.

Toxin based models: Several neurotoxins have been used to induce dopaminergic neurodegeneration such as 6-hydroxy dopamine (6-OHDA), paraquat, 1-methyl-4-phenyl-1,2,3,6-tetrahydropyridine (MPTP) and rotenone (Hirsch et al., 2003; Hirsch, 2006). Rotenone and MPTP share a similarity in their ability to potentially inhibit complex I, however all of these trigger ROS generation.

i. MPTP (1-Methyl-4-phenyl-1, 2,3,6-tetrahydropyridine) mouse model of Parkinson's disease: MPTP was discovered as a model for PD in 1982 when young drug abusers after taking intravenous injection of a synthetic heroine substitute, 1-methyl-4-phenyl-4-propionoxypiperidine (MPPP), an analog of the narcotic meperidine (Demerol), developed a rapidly progressive Parkinsonian syndrome (Langston et al., 1983). It was later realized that MPTP, an accidental byproduct during synthesis of MPPP, was present as a neurotoxic contaminant. MPTP induces an irreversible and severe syndrome characterized by all features of Parkinson's disease in humans and monkeys, including tremor, rigidity, slowing of movement, postural instability, and freezing. In MPTP treated non-human primates beneficial response to levodopa and development of long-term motor complications to medical therapy are nearly identical to that seen in PD patients. The susceptibility to MPTP increases with age in both monkeys and mice (Irwin et al., 1993; Rose et al., 1993; Ovadia et al., 1995). Since its discovery, MPTP mouse model has been used to understand the mechanism of selective degeneration of SNpc dopaminergic neurons. Although mice do not display clinical features of PD, the pattern of dopaminergic cell loss in MPTP treated mice is similar to that seen in PD (Seniuk et al., 1990; Muthane et al., 1994), thereby establishing MPTP as a widely used model.

Being highly lipophilic, MPTP, soon after systemic administration, crosses the blood-brain barrier (Markey et al., 1984) where it is oxidized to 1-methyl-4-phenyl-2, 3-dihydropyridinium ($MPDP^+$) by monoamine oxidase B (MAO-B) in glia and serotonergic neurons and then further converted to the active toxic component, MPP^+ (1-methyl-4-phenylpyridinium) by spontaneous oxidation and released into the extracellular space. MPP^+ being a high-affinity substrate for the dopamine transporter (DAT; (Javitch et al., 1985; Mayer et al., 1986)), enters the neurons and gets sequestered within the mitochondria on account of mitochondrial transmembrane potential (Ramsay and Singer, 1986). MPP^+ can also be taken up by vesicular monoamine transporter 2 (VMAT2) and its sequestration in vesicles can provide protection against subsequent neurotoxicity (Staal and Sonsalla, 2000).

Mode of action of MPTP: On entering mitochondria, MPP^+ impairs oxidative phosphorylation by inhibiting mitochondrial complex I (Nicklas et al., 1985) thus leading to rapid depletion of tissue ATP, particularly in the striatum and ventral midbrain (Chan et al., 1991; Fabre et al., 1999), regions most vulnerable to MPTP.

Generation of oxidative stress is an early event downstream to inhibition of complex I. ROS production ensues due to obstruction in flow of electrons through complex I by MPP^+; free radicals especially superoxide ($O_2^{\cdot-}$) are also generated in redox cycling of

MPP$^+$ (Hasegawa et al., 1990; Hasegawa et al., 1997). NADPH cytochrome P450 reductase reduces MPP$^+$ to MPP· radical (Sinha et al., 1986) which then reduces oxygen to form a superoxide anion regenerating MPP$^+$ back for the redox cycle. Further, superoxide radical leads to generation of hydroxyl radical by the subsequent action of superoxide dismutase and catalase in the presence of iron.

$$MPP^+ \rightarrow (MPP·)$$
$$(MPP·) + O_2 \rightarrow O_2^{-·} + MPP^+$$
$$O_2^{-·} + 2H^+ \rightarrow H_2O_2$$
$$H_2O_2 \rightarrow OH· + OH^-$$
$$(MPP·) + H_2O_2 \rightarrow MPP^+ + OH· + OH^-$$

Findings from *in vivo* studies support the premise for ROS induced damage in MPTP-induced neurodegeneration. Mice transgenic for superoxide dismutase-1 (SOD1), a key ROS scavenging enzyme, are resistant to MPTP-induced dopaminergic neuron degeneration (Przedborski et al., 1992), thus implying a key role for reactive species, including NO (nitric oxide), as critical effectors in MPTP toxicity (Przedborski and Vila, 2003). Previous studies from our laboratory have demonstrated that single dose of MPTP (30mg/kg body weight, s.c) causes mitochondrial complex I inhibition in striatum and midbrain which can be reversed *in vitro* by dithiothreitol indicating that protein thiol oxidation is the primary cause for complex I inhibition (Annepu and Ravindranath, 2000). After exposure to a single dose of MPTP (30mg/kg body weight, s.c), complex I activity in the midbrain and striatum were affected for up to 24 hours and thereafter recovered without showing any signs of permanent damage. Only chronic treatment with MPTP causes irreversible dopaminergic cell loss (Sriram et al., 1997). Since MPTP is clearly linked to a form of human Parkinsonism and is the most widely studied model drug for PD, we, henceforth, used this model for our study. Our lab has also previously demonstrated that female mice are resistant to MPTP toxicity and do not show complex I inhibition on MPTP administration. Hence, we used MPTP as our toxin model to examine the altered signaling in male and female mice.

ii. **Paraquat:** N,N' dimethyl-4,4 bipiridinium, (paraquat) is a herbicide having structure similar to MPP$^+$ (N-methyl-pyridinium group at the second carbon instead of the phenyl group in MPP$^+$) which induces PD like pathogenesis. Toxicity is mediated by generation of ROS. Administration of paraquat to mice results in degeneration of dopaminergic neurons in SNpc, decrease in levels of striatal dopamine, motor impairments and formation of α-synuclein containing inclusion bodies (McCormack et al., 2002). It can be administered systemically; however it generates global toxicity in the organism.

iii. **Rotenone:** Rotenone, a commonly used insecticide, being lipophillic, enters cell membrane easily. Its accumulation in subcellular organelles, such as mitochondria impairs oxidative phosphorylation by inhibiting mitochondrial complex I (Schuler and Casida, 2001). It can be administered systemically and its low-dose chronic administration to rats results in formation of inclusions bodies containing α-synuclein in substantia nigra (Betarbet et al., 2000). Rotenone results in depletion of striatal dopamine levels, forms LB like aggregates and causes massive loss of dopaminergic neurons. Since it causes global toxicity in the organism and not the selective degeneration of dopaminergic neurons of only substantia nigra pars compacta (SNpc), it is not a preferred candidate as a model neurotoxin for PD.

iv. **6-Hydroxy dopamine (6-OHDA):** 6-Hydroxy dopamine has been used as a toxin model for PD in both rodents and non human primates. 6-OHDA exerts its toxicity by generating ROS, reducing striatal dopamine levels and causing massive loss of dopaminergic neurons (Glinka and Youdim, 1995). It does not form intracellular aggregates but alters calcium homeostasis in mitochondria (Frei and Richter, 1986). The dopaminergic and noradrenergic transporters preferentially take up 6-OHDA causing damage to monoaminergic neurons (Luthman et al., 1989). Since 6-OHDA cannot cross the blood brain barrier, it is administered by local stereotaxic injection into SN or striatum to target the nigrastriatal dopaminergic pathway. Since it causes massive and fast neurodegeneration non-selectively and does not mimic the PD pathology observed in humans, i.e. selective degeneration of dopaminergic neurons in SNpc but not ventral tegmental area (VTA), it is also not a preferred model neurotoxin. Only MPTP reproducibly recapitulates the features of human PD pathology and hence has been used extensively to understand the mechanisms underlying the pathogenesis.

3.0 Cell death in PD: Parkinson's disease is characterized by slow and progressive demise of selected population of neurons including dopaminergic neurons of SNpc, selected aminergic brain-stem nuclei (both catecholaminergic and serotoninergic), the cholinergic nucleus basalis of Meynert, hypothalamic neurons, and small cortical neurons, as well as neurons in olfactory bulb, sympathetic ganglia, and parasympathetic neurons in the gut. Susceptibility of dopaminergic neurons also varies in different projection areas and within SNpc, neuronal loss tends to be greatest in the ventrolateral tier (loss is estimated to be 60 to 70 percent at the onset of symptoms), followed by the medial ventral tier and dorsal tier (Fearnley and Lees, 1991). Cell loss results in a regional loss of striatal dopamine, most prominently in the dorsal and intermediate subdivisions of the putamen (Kish et al., 1988), a process that is believed to account for akinesia and rigidity. This pattern of cell loss also correlates with the degree of expression of messenger RNA for the dopamine transporter (Uhl et al., 1994). It has also been suggested that a greater degree of medial nigral cell loss with projections to the caudate nucleus could result in more cognitive deficit (Gibb and Lees, 1991). Other possible clinical pathological correlations include neurodegeneration in olfactory bulb causing anosmia, degeneration in the intermediolateral columns of the spinal cord, sympathetic and parasympathetic ganglia and possibly the amygdaloid nucleus causing behavioral dysfunction including depression, which occurs in approximately one quarter of patients (Braak et al., 1994; Tandberg et al., 1996; Sawamoto et al., 2008). Inspite of its heterogenous pathology, appearance of Lewy bodies, the loss of dopaminergic neurons in SNpc and depletion of striatal dopamine predominate symptomatically, in most patients, in the early years.

3.1 Cell death pathways underlying pathogenesis of PD: PD is a chronic progressive neurodegenerative disorder with no proven cure so far. There is an urgent need to develop therapies which can either retard or stop the progression of the disease or reverse the neurodegeneration associated with it. Current therapies based on dopamine replacement are mainly symptomatic and therefore, it becomes very important to understand the molecular and biochemical pathways underlying the pathogenesis of PD. Information obtained through genetic approach has facilitated the understanding of signaling pathways that go errant during the progression of the disease. Understanding the interaction of signaling pathways would help to identify crucial points which could prove to be potential drug targets.

3.2 Mechanisms of neurodegeneration in PD: Prevalence of neurodegenerative disorders including PD is on rise rapidly. Alhough, the quest for therapeutic interventions is on rise, yet, most of the clinical trials have been unsuccessful indicating that we are missing out on some critical issue regarding the therapeutic entry point. Adding to complexity is the fact that, neuronal dysfunction/loss is a consequence of several destructive phenomena occurring within the affected cell or an organism as a whole. It could either be aggregation of abnormal protein assemblies which may trigger vicious aberrant signaling cascades leading to the selective demise of the cells, or it may be the impairment of energy metabolism due to failure of respiratory chain thus leading to oxidative stress, mitochondrial dysfunction which may further propagate cell death signaling. Thus, cell death in PD is a fine interplay among protein aggregation, mitochondrial dysfunction and oxidative stress, which contribute to the degeneration.

3.3 Protein aggregation and Ubiquitin Proteasome System (UPS): Majority of neurodegenerative diseases are characterized by protein aggregation and are termed proteinopathies (Jellinger, 2009). These protein aggregates may be cytosolic (PD, HD), intranuclear (spinocerebellar ataxia) or secreted as amyloid-β oligomers which form extracellular aggregates (AD) (Jellinger, 2008). PD is pathologically characterized by presence of Lewy bodies (LBs), regarded as its hallmark. Protein aggregation in PD generally results from abnormal expression levels of certain proteins such as α-synuclein, Parkin, UCH-L1 and DJ-1 (Barrachina et al., 2006). Protein encoding parkin has UBiquitin Like (UBL) domain (N-terminal) which serves as proteasome binding motif and two RING finger motifs (C-terminal), which recruit the E2 component of the ubiquitin machinery (Shimura et al., 2000). Most of the point mutations in parkin reside in its RING-IBR-(In Between RINGS)-RING domain, thus causing its inactivation and lead to the accumulation of its substrates such as Cell Division Control related protein (CDCrel-1), Parkin associated endothelial like (Pael) receptor and α-synuclein which upon aggregation elicit cell death via Unfolded Protein Response (UPR), a mechanism which triggers an endoplasmic reticulum stress response. Mutations in UCH-L1 gene also lead to defect in UPS, since it helps in ubiquitination of proteins including α-synuclein. Mutations in UCH-L1 also partially inactivate the enzyme thus causing shortage of free ubiquitin which leads to general impairment of UPS.

Third important player which links impairment of UPS to PD pathogenesis is α-synuclein (PARK1). The wild type protein is monomeric, which oligomerizes to β-pleated sheets at higher concentration and forms the neuronal intracytoplasmic LBs in PD.

Mutations in N-terminal domain of α-synuclein (A30P & A53T) increase the propensity of its fibrillization. Both monomeric as well as aggregated form of α-synuclein bind to S6 proteasome subunit and inhibit proteasomal function (Snyder and Wolozin, 2004). Besides this, mutation in DJ-1 gene (L166P), a point mutation, destabilizes the protein thus promoting its rapid proteasome mediated degradation (Miller et al., 2003) indicating that the loss of function leads to the PD related pathogenesis. In sporadic PD, there is age related tendency to accumulate oxidatively damaged protein and failure of UPS to remove misfolded protein may be the underlying cause of pathogenesis. There are reports which suggest that α but not β subunits of the core catalytic 20S subcomplex of 26S proteasome are lost (McNaught et al., 2002) and there is also impairment of 20S proteasomal enzymatic activity in substantia nigra (SN) (McNaught and Jenner, 2001). Proteasome activator 19S/PA700 as well as the levels of PA28 regulator was found to be decreased in SN.

3.4 Mitochondrial dysfunction and oxidative stress: Mitochondria, in spite of being the primary source of energy production in a cell can also contribute to neurodegeneration through accumulation of mitochondrial DNA (mtDNA) mutations leading to impairment of electron transport chain (ETC) and generation of reactive oxygen species (ROS). Mutations in mtDNA result in decreased respiratory enzyme activity and ATP production (Van Houten et al., 2006).

Mitochondria also produce ROS which contributes to neurodegeneration. Electron carriers in ETC of mitochondria are efficiently capable of ROS production which is normally neutralized by extensive network of defensive antioxidant proteins, such as superoxide dismutase (SOD), glutathione peroxidase (Gpx), glutathione reductase (GR) and catalase. Oxidative insult can cause an imbalance in the ROS generation and its clearance thus leading to net production of ROS which is deleterious and causes mitochondrial dysfunction.

Oxidative stress is implicated in most of the neurodegenerative disorders such as AD, PD and ALS and all these diseases also exhibit ETC impairment and cellular damage mediated by free radical (Adam-Vizi, 2005). ROS potentially are extremely harmful to the cells and cause oxidative damage to lipid, proteins and DNA which results in inhibition of mitochondrial transcription, in turn, impairing ETC, ATP production and further generation of ROS in a vicious circle. Such progressive depletion of ATP may result in loss of Na/K-ATPase activity leading to membrane depolarization, loss of cellular homeostasis, opening of mitochondrial permeability transition pore and propagation of apoptosis. Cells harbor a complex, interdependent efficient antioxidant defense system to deal with free radicals. These include vitamins (A, C and E), trace elements (Se, Zn and Mn) and compounds like reduced glutathione (GSH), ubiquinone, NADPH, coenzyme Q and antioxidant enzymes which are reportedly induced during oxidative conditions. However, impairment of antioxidant defense system during diseased conditions may render brain more vulnerable to oxidative damage.

Mitochondrial dysfunction was first implicated in PD when MPP^+, the active metabolite of MPTP, was found to inhibit complex I of mitochondrial electron transport chain (ETC) in drug abusers resulting in clinical and pathological phenotype of Parkinsonism. Later, it was also observed in naturally occurring idiopathic PD, where glutathione depletion and complex I deficiency was found in SN of patients (Lee et al., 2009). Further most of the nuclear genes associated with PD such as α-synuclein, Parkin, DJ-1, PINK1, LRKK2 and HTRA2 implicate mitochondria in pathogenesis due to their localization and function in mitochondria.

Mutations in α-synuclein, a major component of LBs, increases the propensity for formation of fibrillar aggregates and a close relationship exists between abnormal protein accumulation or degradation, oxidative stress and mitochondrial dysfunction. Parkin deficiency or mutations in it also lead to oxidative stress and mitochondrial dysfunction and impairment in complex I activity. DJ-1, another multifunctional gene product, which is known to interact with α-synuclein, Parkin and PINK1, protect against cell death especially induced by oxidative stress. In Drosophila, DJ-1 undergoes progressive oxidative inactivation during ageing thus increasing sensitivity to oxidative stress thereby partially explaining age dependence of sporadic PD (Meulener et al., 2006). Deficiency of PINK1, a mitochondrial kinase, in Drosophila, leads to mitochondrial impairment, increased sensitivity to paraquat and rotenone and causes degeneration of flight muscles and dopaminergic neurons and this pathology can be rescued by overexpressing Parkin (Gegg et al., 2009; Venderova et al., 2009). HTRA2, the most recently PD associated gene is a quality control agent within mitochondria. It helps in maintaining mitochondrial function and knockout mice develop striatal degeneration and Parkinsonism. Mutations in HTRA2 impair its protease activity causing mitochondrial swelling and decreases mitochondrial membrane potential (MMP) (Strauss et al., 2005; Bogaerts et al., 2008).

3.5 Modes of cell death: PD is a disorder with multiple pathways of cell death operating during its progression. It is not clear whether these pathways converge to one central cell death mechanism, or they are independently and distinctly involved in different forms of genetic and sporadic PD. Neurodegeneration in PD is associated with aggregation of misfolded proteins, generation of reactive oxygen or nitrogen species leading to oxidative stress, complex I inhibition (Schapira et al., 1989), mitochondrial dysfunction and calcium entry leading to excitotoxicity. These phenomenons may lead to cell death through classic apoptotic pathways, endoplasmic reticulum (ER) stress, as well as neuronal nitric oxide synthase (nNOS) activation, DNA damage, and poly (ADP-ribose) polymerase (PARP) activation. Modes of cell death can be classified as discussed below.

i. **Programmed cell death (PCD) in PD:** Biochemical activation of classical apoptosis occurs through two pathways. Extrinsic pathway includes activation of death receptors at the cell surface, such as Fas which lead to the activation of caspase 8 or 10. Intrinsic pathway is triggered by release of cytochrome c from mitochondria and is associated with activation of caspase-9. Intrinsic pathway may also originate from endoplasmic reticulum and activate caspase-9. Formation of misfolded proteins, a hallmark for neurodegenerative disorders including PD, induces cell death via mediators like BAX/BAK, p53, PUMA and NOXA, which are all Bcl-2 family proteins. Presence of more number of TUNEL positive dopaminergic neurons in brain samples of PD patients support occurrence of apoptosis in PD (Mochizuki et al., 1996). Immunolocalization of Bax and activity of effector protease caspase 3 has been found to be more in brains (SNpc) of PD patients when compared to controls thus emphasizing on the presence of more pro-PCD proteins in the affected brains (Hartmann et al., 2000; Tatton, 2000; Hartmann et al., 2002). MPTP induces activation of p53 and subsequent upregulation of Bax. Inhibition of p53 by chemical means or by its genetic ablation attenuates MPTP induced Bax upregulation and degeneration of dopaminergic neurons (Trimmer et al., 1996; Duan et al., 2002). Jun C-terminal kinase (JNK) pathway is also reportedly induced by MPTP. Further, blockade of intrinsic PCD pathways by intrastiatal injections of adeno associated viral (AAV) vector harboring a dominant negative form of Apaf1 prevents the MPTP induced activation of caspase-3 and ensuing SNpc neuronal death.

ii. **Autophagy in PD:** Autophagy in PD is neuroprotective. It complements proteasomal pathways in clearance of aggregated protein, damaged mitochondria by an intracellular phagocytotic process by forming autophagosomes which fuse with lysosomes for hydrolytic degradation of its contents. As per recent reports autophagic-lysosomal dysfunction may also be a causative mechanism involved in PD. Marked accumulation of autophagic vacuoles and impairment of lysosomal system and UPS has been noticed in cells overexpressing wild-type and A53T mutant α-synuclein (Stefanis et al., 2001) since clearance of mutant α-synuclein is strongly dependent on both UPS and macroautophagy (Webb et al., 2003). Mice deficient in Atg7 and Atg5, specifically in the CNS exhibit behavioural deficits. Histological data analysis depicts loss of cerebral and cerebellar neurons and presence of ubiquitinated proteins abundantly in the neurons although UPS was intact (Bjorkoy et al., 2005; Hara et al., 2006).

iii. **Cell death by necrosis:** The decision of the cell to undergo apoptosis or necrosis depends upon the abundance of intracellular energy stores. Whereas apoptosis requires a minimum amount of intracellular ATP, necrosis is accompanied by its total depletion. Many cellular cytotoxins such as hydrogen peroxide and other oxidants induce PCD at low concentration and necrotic cell death at higher concentration, thus inducing something called aponecrosis (Formigli et al., 2000). This happens presumably because the homeostatic processes in the cell are overwhelmed before cell death programmes are completely executed.

4.0 Role of mitochondria in PD: The first evidence of mitochondrial role in human disease emerged in 1958 when a Swedish patient was identified with symptoms of severe polydipsia, polyphagia, weight loss, and weakness (Luft et al., 1962). Biochemical analysis revealed that the patient had a partially uncoupled respiration. The disease was associated with spontaneous release of mitochondrial calcium, which may lead to abnormal calcium cycling and thereby sustained stimulation of respiration and loose coupling (Luft, 1997). The discovery of Luft's disease unveiled a previously unnoticed role of mitochondria in human disease and initiated explorations in the field of mitochondrial pathology. Since then, several human mitochondrial neurodegenerative diseases have been discovered (Luft et al., 1962) which involve selective population of neurons in the CNS. These diseases have been associated with specific, inherited mitochondrial DNA mutations which also accompany deficiency in electron transport chain (ETC). Despite the obvious differences in the primary etiologies of neurodegenerative disorders, a role for mitochondrial dysfunction has been postulated in the pathogenesis of these diseases.

4.1 Mitochondrial inheritance and genetics in PD: Mitochondria possess their own DNA (mtDNA), which is inherited through the maternal line. The mtDNA encodes 13 polypeptides, all of which are components of the respiratory chain, and also encodes a complement of rRNAs and tRNAs, necessary for intra organelle protein synthesis (Taanman, 1999). Although nuclear DNA encodes most mitochondrial components, mtDNA defects can cause numerous diseases, many of which are associated with neuronal degeneration. Numerous findings have lead to realization of possible mitochondrial involvement in the pathogenesis of PD. Complex I deficiency and oxidative damage have been observed in SN of PD patients (Schapira et al., 1990), along with reduced immunoreactivity for complex I subunits (Hattori et al., 1991). Complex I is an important source of free radicals in the cell (Lenaz, 1998), hence abnormalities in mitochondrial complex I, as observed in platelets and muscle of Parkinson's disease patients (Parker et al., 1989; Mizuno et al., 1995), could be responsible for increased lipid peroxidation and DNA damage found in PD brains (Dexter et al., 1994). Cybrids containing mtDNA from PD platelets also showed reduced complex I activity (Gu et al., 1998), strongly suggesting that inherited and/or somatic mtDNA mutations might be responsible for the biochemical phenotype in PD. Polymorphism in NDUFV2 complex I subunits (24 kDa) were found in some patients with PD (Hattori et al., 1998). In certain cases, these mutations might represent the primary cause of the disease, as maternally inherited forms of PD or Parkinsonism with complex I deficiency have been reported (Simon et al., 1999). More frequently, mtDNA mutations, either inherited or acquired could contribute together with nuclear gene mutations and environmental factors to the pathogenesis of PD.

4.2 Oxidative stress and Mitochondria: Oxidative stress refers to the imbalance which favors production of ROS over the ability of a cell's antioxidant defenses. This imbalance causes an accumulation of oxidatively damaged molecules, which results in cellular dysfunction. ROS is primarily generated within mitochondria thus leading to mitochondrial dysfunction which further causes decreased ATP production and other potentially detrimental effects, such as impaired intracellular calcium buffering and further generation of ROS. Mitochondria essentially also activate certain forms of apoptosis (Tatton and Olanow, 1999).

i. **Mitochondria as the main source of ROS:** Mitochondria are the "power house of the cell" where metabolites are converted into ATP through oxidative phosphorylation. Majority of ROS is produced by mitochondria itself and it plays a regulatory role in cellular metabolic processes by activating enzymatic cascades and specific redox-sensitive signaling pathways. Approximately 1-2% of molecular oxygen consumed during normal physiological respiration is converted into superoxide radicals. One electron reduction of molecular oxygen results in formation of superoxide anion (O_2^-) which is a relatively stable intermediate and is regarded as the precursor of most ROS. This one electron reduction of molecular oxygen is thermodynamically favored and there are several redox centres in mitochondrial electron transport chain (ETC) which may leak electrons to molecular oxygen serving as the primary source of superoxide production.

ii. Electron Transport Chain (ETC) dysfunction and generation of ROS by complex I & III: Primary source of free radicals in neurodegeneration is still controversial, however there is mounting evidence that impairment of ETC is integral to these diseases. Free radicals are generated as normal products of cellular aerobic metabolism. Superoxide (O_2^-) and hydroxyl ($OH^.$) ions are the most predominant cellular free radicals and together with hydrogen peroxide (H_2O_2) and peroxynitrite ($ONOO^-$) constitute reactive oxygen species (Halliwell, 1989b). Besides this, defects in ETC also act as a source of free radical generation by blocking the normal flow of electrons down the chain and ultimate reduction of molecular oxygen to water. Complex I accounts for the bulk of O_2^- production followed by complex III. Ubiquinone (Ub), which is a component of mitochondrial respiratory chain and connects complex I and III plays a key role in production of O_2^- by complex III. Oxidation of Ub proceeds in a set of reactions known as Q-cycle and unstable semiquinone is responsible for formation of O_2^-. Mitochondria being the most powerful intracellular source of ROS has approximately five to ten fold higher concentration of O_2^- when compared to cytosol or nuclear space. Dismutation of superoxide anions by superoxide dismutase results in the production of hydrogen peroxide (H_2O_2) which further reacts with residual O_2^- in a Haber-Weiss reaction or undergoes cleavage driven by Fe^{+2}/Cu^{+2} in Fenton's reaction thus generating highly reactive and toxic hydroxyl radical. The Haber-Weiss cycle consists of the following two reactions:-

$$H_2O_2 + OH^. \rightarrow H_2O + O_2^- + H^+$$
$$H_2O_2 + O_2^- \rightarrow O_2 + OH^- + OH^.$$

Moreover, high iron content is reported in some areas of brain, which further catalyses the generation of $OH^.$. Since iron has a loosely bound electron, it can exist in more than one valency state - the stable redox form being Fe^{3+} and its bivalent form, Fe^{2+}.

The bivalent form is capable of transferring one electron and facilitating free radical generation. The reaction of Fe^{2+} with H_2O_2 produces •OH and is called as Fenton reaction.

$$H_2O_2 + Fe^{2+} \rightarrow Fe^{3+} + OH^{\cdot} + OH^-$$
$$H_2O_2 + Cu^+ \rightarrow Cu^{2+} + OH^{\cdot} + OH^-$$

Besides this, MPP⁺ itself induces ROS generation by inhibiting complex I and ensuing ROS production due to obstruction of flow of electrons through complex I as has been described in earlier section. Further, superoxide radicals leads to generation of hydroxyl radical by the subsequent action of superoxide dismutase, catalase and in presence of iron.

iii. **Mitochondrial targets of ROS:** Oxidative phosphorylation produces ROS continuously at high rate within the brain, which can cause peroxidation of lipids, lead to DNA strand breaks and modify proteins. However, there are mechanisms developed during evolution to detoxify and/or prevent the generation of hydroxyl radical. For instance, removal of H_2O_2 and superoxide prevents the generation of ROS that are formed during iron catalyzed Fenton reaction or by Haber-Weiss reaction (Halliwell, 1989a).

The major enzymatic antioxidants which catalyze the dismutation of superoxide to oxygen and hydrogen peroxide include Cu/Zn- and Mn-superoxide dismutase (SOD). Catalase (CAT) decomposes hydrogen peroxide directly into water and oxygen. Peroxiredoxins and glutathione peroxidase (GSH-Px) detoxifies hydrogen peroxide using reduced Trx and cellular glutathione respectively.

$$O_2^{-\cdot} + O_2^{-\cdot} + 2H^+ \xrightarrow{SOD} H_2O_2 + O_2$$
$$H_2O_2 + H_2O_2 \xrightarrow{Catalase} 2H_2O + O_2$$
$$H_2O_2 + GSH \xrightarrow{GSH\ peroxidase} GSSG + H_2O$$

There are non-enzymatic antioxidants and metal chelators, which include vitamin E, vitamin C, beta-carotene, selenium, ubiquinone, ferritin, ceruloplasmin, and uric acid. Increased free radical or peroxide production results in induction of antioxidant defense mechanism (Cohen, 1994). Higher expression levels of antioxidant defense systems may offer some protection against free radicals but may not completely prevent oxidative damage since the efficiency of gene expression may decline with advancing age. Situation is further deteriorated because brain contains lower amounts of superoxide dismutase, catalase and glutathione peroxidase activity compared to other organs such as liver and kidney (Coyle and Puttfarcken, 1993). This leads to accumulation of oxidized products which serve as markers of excessive oxidative stress during ageing and thus implicate a role of oxidative stress in neurodegenerative disorders. Besides mitochondria, which are main source of generation of reactive oxygen species, external stress such as ultraviolet (UV), ionizing irradiation and drugs also generate ROS which if not neutralized by the cell's antioxidant defense system, impairs mitochondria thus further increasing the production of ROS in a cyclical reaction.

4.3 Targets of ROS (DNA, protein, lipids): ROS generated by mitochondria or elsewhere in the cell can cause damage to cellular macromolecules including nucleic acids, proteins and phospholipids.

i. **Damage to DNA:** Hydroxyl radical can potentially modify purine and pyrimidine bases, deoxyribose backbone, cause single and double strand-breaks in DNA as well as cause cross linking with other molecules. Hydroxylation of deoxyguanosine residues produces 8-hydroxy-2'-deoxyguanosine (8OH-2-dG), which can be used as marker of oxidative damage.

ii. **Lipid peroxidation:** Polyunsaturated fatty acid residues present in phospholipids are extremely sensitive to oxidation and hydroxyl radical generated by Fenton's reaction is the most potent inducer of lipid peroxidation wherein Fe^{2+} acts as a strong catalyst.

iii. **Damage to protein:** Several types of oxidative damage to proteins have been demonstrated such as oxidation of sulfhydryl groups resulting in reversible disulfide cross-linking, reaction with aldehydes, protein-protein crosslinking, addition of oxidative adducts on amino acid residues in close vicinity to metal-binding sites and direct oxidation of amino acids resulting in formation of protein carbonyls. Oxidative stress damages proteins and marks them for degradation by intracellular proteases (Kasai et al., 1986; Stadtman et al., 1993).

4.4 Dopamine as an inducer of reactive oxygen species (ROS): The most striking feature of PD is selective vulnerability of a restricted population of neurons, SNpc dopaminergic neurons being the most vulnerable (Lotharius and Brundin, 2002). Since SNpc neurons exist in an oxidative environment, with high concentrations of iron, neuromelanin and dopamine (DA) and auto-oxidation of DA can lead to the generation of an array of toxic metabolites and ROS, DA to an extent itself contributes to the demise of SNpc neurons. DA metabolism is highly oxidative and produces toxic semiquinone species. It is metabolized through two possible pathways leading to the production of homovanillic acid (HVA). Its metabolism through catechol-O-methyl transferase (COMT) leads to the formation of 3-O-methyldopamine which is further catabolized to HVA through monoamine oxidase (MAO). Metabolism of DA through MAO leads to the formation of 3,4-dihydroxy-phenylacetaldehyde (DOPAL), a highly reactive aldehyde, which is further catabolized by aldehyde dehydrogenase (ALDH) to 3,4-dihydroxy-phenylacetic acid (DOPAC). DOPAL in combination with H_2O_2 (both products of MAO) can lead to the formation of highly toxic hydroxyl radical (OH·) in an iron-catalyzed reaction (Li et al., 2001b). Elevated iron concentrations in SNpc have been implicated in the development of idiopathic PD (Dexter et al., 1989; Hirsch et al., 1991) since it can catalyze the formation of ROS from the oxidation of dopamine or levodopa (Jenner and Olanow, 1998). Moreover, during pathogenesis of PD, levels of antioxidant enzymes and glutathione (GSH) are lowered thus further adding to the severity of the disease (Sofic et al., 1992; Sian et al., 1994b; Sian et al., 1994a). The cause of GSH depletion in PD is unknown and there is no evidence for reduced synthesis (Sian et al., 1994b). It may be due to increased oxidative stress secondary to the complex I defect or to levodopa therapy itself (Jenner and Olanow, 1998).

4.5 Kreb cycle dysfunction and inactivation of Fe-sulfur proteins: Superoxide anion O_2^- can directly cause oxidative inactivation of iron-sulfur (Fe-S) proteins, such as aconitases, thus releasing iron which further induces the production of hydroxyl radical. Mitochondrial aconitases catalyze the conversion of citrate to isocitrate in Kreb's cycle. Even partial inhibition of aconitase by oxidative inactivation results in Kreb's cycle dysfunction thus impairing the energy production. Some other Fe-S proteins are also affected by O_2^- such as complex I (NADH dehydrogenase) and complex II (succinate dehydrogenase) thus impairing the energy metabolism completely.

4.6 Effect of ROS on Ca^{2+} dependent mitochondrial permeability transition (MPT): Apart from ATP production in aerobic cells, mitochondria also play a role in regulation of intracellular Ca^{2+} ion homeostasis. ROS production in mitochondria can potentially terminate into Ca^{2+} dependent MPT, which plays a key role in certain modes of cell death. Mitochondria can take up and sequester Ca^{2+} until it exceeds the retention threshold concentration for Ca^{2+}, following which it is effluxed from mitochondria through a proteinaceous channel called permeability transition pore (PTP). The threshold for Ca^{2+} retention to cause MPT lowers when sequestration of Ca^{2+} is also accompanied by oxidative stress and depletion of adenine nucleotides. Thus, oxidative stress generated by mitochondrial ROS along with perturbed Ca^{2+} homeostasis sensitizes it towards MPT induction and contributes to subsequent cellular damage. MPT further causes mitochondrial dysfunction which can lead to either necrosis if there is complete ATP depletion or to caspase dependent apoptosis if MPT induction occurs in a subset of mitochondria within a cell, while the rest of the organelles are still able to maintain mitochondrial membrane potential (MMP) and generate ATP.

5.0 Mitochondrial membrane permeability transition (MPT): Under conditions of oxidative stress, there is the sudden increase in Ca^{2+} dependent permeability of inner mitochondrial membrane (IMM) resulting in loss of MMP ($\Delta\Psi_m$), mitochondrial swelling and rupture of outer mitochondrial membrane (OMM) thus causing mitochondrial membrane permeability transition (MPT). MPT occurs due to the opening of permeability transition pore complex (PTPC) comprising voltage dependent anion channel (VDAC; outer membrane channel), adenine nucleotide translocator (ANT; inner membrane channel) and cyclophilin D (Cyp D; in mitochondrial matrix). The above said components of putative PTP are considered to play a role in MPT. VDAC is also known to play a role in MPT since polyclonal antibodies recognizing different epitopes of the channel could inhibit VDAC activity as well as Ca^{2+} induced MPT (Shimizu et al., 2001). ANT is also considered important for inducing MPT but mitochondria from ANT knockout mice went through MPT, even though the Ca^{2+} threshold was higher (Kokoszka et al., 2004). Cyp D deficient mitochondria do not undergo cyclosporine A (CycA) sensitive MPT triggered by a variety of MPT inducers such as Ca^{2+} and H_2O_2. Hence, on induction of apoptosis VDAC channels increase their conductance leading to MPT, MMP and eventual cell death, CypD can act as a "PTPC organizing centre by providing other mitochondrial proteins a docking site in the matix or internal membrane and it also causes conformational changes in ANT which triggers MPT. Thus all three of them have a role to play in MPT.

5.1 Maintenance of MMP by electron transport chain (ETC): Normally in a healthy cell under physiological conditions, the inner mitochondrial membrane is almost impermeable to ions including protons, which permits complex I-IV of ETC to build up a proton gradient across inner mitochondrial membrane (Mitchell and Moyle, 1965). The generation of electrochemical gradient across inner mitochondrial membrane results in charge imbalance and forms the basis of inner mitochondrial transmembrane potential ($\Delta\Psi_m$). This proton gradient is eventually utilized by complex V of ETC for ATP generation (Mitchell and Moyle, 1965). Hence influx and efflux of metabolites across inner mitochondrial membrane occurs in a tightly regulated fashion through selective channels and transport proteins. Similarly, permeability of OMM is also tightly regulated and only flow of solutes less than 5 kDa is possible through VDAC present in outer mitochondrial membrane.

5.2 Permeabilization of outer mitochondrial membrane (OMM): Mitochondrial membrane permeabilization can be initiated by changes in permeability transition pore complex (PTPC) which acts as a supra-molecular channel between inner and outer mitochondrial membrane. Under physiological conditions PTPC has a 'flickering' state characterized by reduced conductance, allowing passage of metabolites between mitochondrial matrix and cytosol. During oxidative conditions, such as ROS generation or in response to Ca^{2+} overload, the PTPC transforms into a high-conductance conformation, thus permitting the unregulated entry of solutes and water into the mitochondrial matrix (driven by osmotic forces) leading to membrane permeability transition (MPT). MPT is characterized by instantaneous dissipation of the mitochondrial transmembrane potential, leading to influx of water molecules and progressive swelling of the mitochondrial matrix. This further leads to outer mitochondrial membrane breakdown and release of intermembrane space proteins into the cytosol (Kroemer et al., 2007) eventually leading to trigger apoptotic cascade in a stepwise manner.

5.3 Mechanism of permeabilization: Mitochondrial membrane permeabilization can result from two distinct but partially overlapping mechanisms. It can be either BAX/BAK mediated mitochondrial outer membrane permeabilization or VDAC/PTPC mediated permeabilization.

i. **Bax/Bak mediated permeabilization:** Under healthy physiological conditions, inactive BAX (BCL-2-associated protein X) and BAK (BCL-2 antagonist/killer) are present in cytosol and remain weakly associated with OMM. During pro-apoptotic conditions, BAX and BAK undergo conformational modifications and get completely inserted into the OMM, thereby creating protein-permeable channels, resulting in a phenomenon called mitochondrial outer membrane permeabilization (MOMP). The protein permeable channel thus formed allows cytosolic leakage of intermembrane space proteins followed by oxidative phosphorylation uncoupling, ROS generation and (eventually) mitochondrial transmembrane potential dissipation (Kroemer et al., 2007). Besides this, BAX/BAK is also known to interact with PTPC protein, VDAC and initiate MOMP (Shimizu et al., 2000).

ii. **VDAC/PTPC mediated permeabilization:** Mitochondrial membrane permeabilization can also be initiated by permeability transition pore complex (PTPC). It is a multi-molecular channel connecting outer and inner mitochondrial membranes and comprises of VDAC, ANT and Cyp D. During ROS generation and increased Ca^{2+} load, PTPC undergoes conformational changes thereby causing a dysregulated and unrestricted influx of water and solutes within the mitochondria as has been described above. According to another model, apoptotic induction favors closed conformation of VDAC which eventually triggers outer mitochondrial membrane rupture (Vander Heiden et al., 1999; Vander Heiden et al., 2000; Vander Heiden et al., 2001).

5.4 Excitotoxicity: Excitotoxicity describes neurotoxic effects of excitatory amino acids, such as glutamate and aspartate, which damage neurons by depolarizing them and causing their death due to over excitation (Olney, 1969; Olney et al., 1971). It plays a role in neuronal death in stroke and neurodegenerative diseases, such as HD, PD and ALS (Simonian and Coyle, 1996). Glutamate is the principal excitatory neurotransmitter in the brain, and it interacts with different receptors thus mediating several neurologic functions such as cognition, memory and movement (Gasic and Hollmann, 1992). It interacts with both postsynaptic ionotropic glutamate receptors [iGluRs] - the N-methyl-D-aspartate (NMDA), kainate and α-amino-3-hydroxy-5-methylisoxazole-4-propionate (AMPA) receptors as well as with metabotropic receptors (Lipton and Rosenberg, 1994). The iGluRs are important in mediating excitotoxicity and neuronal death (Choi et al., 1987) since glutamate, upon binding to them causes depolarization of neurons by allowing Na^+ to enter the cell, which in turn causes opening of the voltage-dependent Ca^{2+} channel and influx of Ca^{2+}. Besides this, NMDA receptor activation allows direct influx of calcium into the cell. Over stimulation of this receptor is also a mechanism for calcium overload in neurons with consequent calcium-mediated neuronal injury (Lipton and Rosenberg, 1994; Orrenius et al., 1996). The pathologic accumulation of glutamate occurs in a number of ways including impaired function of the glutamate uptake transporters, glutamate release from injured neurons and astrocytes, and enhanced presynaptic vesicle release. This further activates NMDA receptors (Lipton and Rosenberg, 1994) leading to excessive influx of Ca^{2+} (Orrenius et al., 1996). The increased concentration of intracellular Ca^{2+} in turn mediates the lethal effects of NMDA receptor over-activation, leading to activity of protein kinase C, phospholipase A2, proteases, and protein phosphatases (Choi et al.,

1988; Lipton and Rosenberg, 1994). These processes, in addition to the activation of proteases result in neuronal death.

i. **Mitochondrial Dysfunction and Excitotoxicity:** Impaired energy metabolism reduces the threshold for excitotoxicity (Novelli et al., 1988). Depletion of intracellular ATP levels using complex IV inhibitors or by glucose deprivation lowers the threshold concentration of extracellular glutamate. Further, inhibition of the mitochondrial respiratory chain or glycolysis also exacerbates the neurotoxic effects of glutamate (Zeevalk and Nicklas, 1991; Simpson and Isacson, 1993) due to reduction in ATP production, which is crucial for maintaining the normal resting potential of the cell membrane via ion pump Na^+/K^+ ATPase. Besides this, partial neuronal depolarization induced by inhibitors of glycolysis or OXPHOS lead to NMDA receptor activation and cell death without an increase in extracellular glutamate concentration (Zeevalk and Nicklas, 1991). NMDA receptor activation also results in increased free-radical production within mitochondria (Reynolds and Hastings, 1995). The increased ROS produced by the mitochondrial respiratory chain as a consequence of NMDA receptor activation depletes ATP levels further and enhances membrane depolarisation and a persistent cycle of energy impairment and NMDA induced excitotoxicity ultimately leads to cell death, thus inferring that mitochondria play a pivotal role in the pathogenesis of cell death due to excitotoxicity (Loeffler and Kroemer, 2000; Zamzami and Kroemer, 2001).

ii. **β-N-Oxalylamino-L-alanine (L-BOAA) & Neurolathyrism:** L-BOAA termed as β-N-oxalyl amino-l-alanine is a glutamate analog, also called as β-N-oxalyl-α,β-diaminopropionic acid (β-ODAP). It is a naturally occurring non-protein amino acid isolated from the chickling pea obtained from the plant *Lathyrus sativus* (Rao et al., 1964). The plant is drought resistant and is grown in the drought prone areas of Africa and Asia. During famine, ingestion of the chickling pea as a staple diet results in a progressive, neurodegenerative condition known as neurolathyrism, a form of motor neuron disease (Selye, 1957) characterized by spastic paraparesis that predominantly targets Betz cells and the corticospinal tracts (Ludolph and Spencer, 1996), predominantly affecting males (Hugon et al., 1988). The clinical pathology involves degeneration of upper motor neurons, anterior horn cells and loss of axons in the pyramidal tracts in the lumbar spinal cord (Cohn and Streifler, 1981).

iii. **Mode of action of L-BOAA:** L-BOAA has high affinity for α-amino-3-hydroxy-5-methyl-4-isoxazole propionic acid (AMPA) receptors (Bridges et al., 1988) and specific AMPA receptor antagonist, NBQX (2,3-dihydroxy-6-nitro-7-sulfamoyl-benzo(F) quinoxaline) prevents L-BOAA-induced neuronal damage, establishing that L-BOAA exerts its action through AMPA receptors (Pearson and Nunn, 1981). L-BOAA is a potent agonist at non-NMDA glutamate receptor sites (Pearson and Nunn, 1981) and facilitates spontaneous and stimulus-evoked release of glutamate from presynaptic elements (Gannon and Terrian, 1989). Exposure to L-BOAA causes mitochondrial dysfunction as evidenced by loss of mitochondrial complex I activity in male mouse brain slices. Complex I inhibition caused by incubating brain slices with L-BOAA could be reversed by disulfide reductants, such as dithiothreitol (DTT), thereby, indicating that L-BOAA oxidatively modifies critical thiol groups of complex I subunits (Sriram et al., 1998).

5.5 Afferent signals from other organelles affecting MPT and cell death effectors: Mitochondria receive incoming pro-apoptotic signals from various organelles and integrate them, thereby occupying a central position in apoptotic signaling. The crosstalk among these signals decides the fate of the cell by favoring either survival or death pathways. Communication among organelles is mediated by several factors such as transcriptional programs, metabolite and ion fluxes, post-translational modifications and redox reactions. The incoming signals may be from nucleus due to DNA damage or from endoplasmic reticulum (ER) due to ER stress induced as a result of defective folding of ER proteins or perturbation in Ca^{2+} gradient built across ER membrane. Mitochondria also receive signals following rupturing of lysosomal membrane and release of cathepsins into the cytosol and initiates apoptosis or necrosis depending on the amount of proteases released. Majority of metabolic activities occur within the cytosol or at the interface of cytosol and mitochondria, thus several metabolites affect MMP and subsequent mitochondrial control of apoptosis. Cytoskeletal components, such as microfilaments and microtubules also regulate survival and demise of cells by modulating motility, polarity, shape maintenance and trafficking of molecules and organelles. Thus all these signals converge and affect mitochondria which then trigger the fate of the cell towards survival or death pathway.

5.6 Mitochondrial cell death effectors: Afferent signals from different sources are received and integrated by mitochondria, which in turn, releases cell death effectors from intermembrane space, if the balance is towards cell death. Amongst the various effectors released, the following are important:-

i. **Cytochrome c:** Release of cytochrome c is among the key events in mitochondrial dependent apoptotic pathway and precedes exposure of phosphatidyl serine and loss of plasma membrane integrity. Upon its release, a fraction of it binds to inositol 1,4,5-trisphosphate receptor (IP_3R) present on ER membrane and amplifies Ca^{2+} dependent and MMP mediated apoptosis. It also promotes assembly of apoptosome leading to the activation of downstream caspase cascade, eventually resulting in apoptosis.

ii. **Smac/DIABLO and Omi/HtrA2:** Second mitochondrial-derived activator of caspase (Smac) resides in intermembrane space of mitochondria and is released into the cytosol following loss of MMP where it neutralizes inhibitors of caspases (IAPs) especially X-linked IAP (XIAP). Similarly, Omi/HtrA2 is a nuclear encoded protease which once released into the cytosol promotes cell death by either antagonizing IAP or by its proteolytic activity.

iii. **Apoptosis inducing factor (AIF):** AIF is a mitochondrial redox active enzyme having pro-apoptotic properties. AIF is normally present in intermembrane space and is required for detoxification of ROS within mitochondria and for maintenance of complex I. Its depletion impairs oxidative phosphorylation and increases vulnerability to oxidative stress. Upon induction of apoptosis, AIF gets released from mitochondria and translocates to the cytosol where it mediates chromatin condensation and DNA fragmentation.

iv. **Endonuclease G (EndoG):** Nuclear DNA fragmentation is a hall mark of apoptosis and is brought about by endonucleases. EndoG is a mitochondrial specific enzyme which is released on induction of apoptosis and like AIF, also translocates to the nucleus and cleaves chromatin DNA into nucleosomal fragments (Li et al., 2001a). Role of caspases in EndoG pathway is however controversial (Parrish et al., 2001; van Loo et al., 2001; Arnoult et al., 2003).

5.7 Mitochondrial permeability transition and neurodegeneration: Mitochondrial dysfunction is implicated in several neurodegenerative disorders, such as PD, AD and HD. Several MMP regulators are also implicated in cell loss following acute neuronal injury. Stimuli like glucose or oxygen deprivation, excessive generation of ROS, Ca^{2+} overload, damage to macromolecules converge to mitochondrial dysfunction and trigger several cascades leading to cell death. In neurodegeneration, majority of neurons die of apoptosis exhibiting the hallmark features of caspase activation, nuclear chromatin condensation and DNA fragmentation and MMP modulators that inhibit mitochondrial apoptosis are able to prevent neuronal loss caused by stroke or hypoglycemic insult (Friberg et al., 1998; Cao et al., 2002). Pathogenesis of chronic neurodegenerative disorders including PD involve loss of MMP and apoptosis as has been observed both in animal models and patients.

Pathologies in these disorders share increased oxidative stress and perturbed energy and ion homeostasis. Moreover several genetic mutation which predispose to PD, invariably modulate MMP and mitochondrial dysfunction. Further, neurotoxins such as MPP$^+$, 6-OHDA and paraquat which induce Parkinsonism also modulate MMP and cause mitochondrial dysfunction.

6.0 Regulation of mitochondrial ROS production: Since mitochondrial respiratory chain is the major source of ROS production, mitochondria are equipped with several antioxidants to neutralize ROS. MnSOD resides in mitochondrial matrix and converts superoxide (O_2^-) to hydrogen peroxide (H_2O_2) which can either diffuse into the cytosol or can be further metabolized by glutathione peroxidase (Gpx) and peroxiredoxin (Prx). In some cell types, CuZnSOD is present in intermembrane space (Okado-Matsumoto and Fridovich, 2001) and converts O_2^- into H_2O_2 which then diffuses into the cytosol. O_2^-, sometimes is also scavenged by cytochrome c or it diffuses into the cytosol through VDAC present in OMM. It may also react with nitric oxide (NO) to form highly reactive peroxynitrite (ONOO$^-$). Hence, mitochondria need constant protection from such reactive species which is provided by enzymatic antioxidant defense system such as ubiquinone, Vitamin E, glutathione (GSH) and GSH linked enzyme antioxidant systems.

6.1 GSH: GSH plays an important role in the detoxification of ROS in brain and since brain has very low levels of catalase and SOD, role of GSH, which maintains thiol homeostasis becomes even more important. Depletion of GSH followed by mitochondrial complex I inhibition is also a concern in PD (Perry et al., 1982; Bharath et al., 2002). Studies carried out in newborn rats with buthionine sulfoximine (BSO), an inhibitor of γ-glutamyl cysteine synthetase (γ-GCS), induced GSH deficiency which caused damage to brain mitochondria (Jain et al., 1991). Moreover, depletion of GSH content in brain also exacerbated toxic effects of MPP$^+$ and elevated ROS production in ischemic conditions

(Mizui et al., 1992; Wullner et al., 1996) thus revealing the significance of maintenance of thiol homeostasis during stress for cell survival. Further, chronic depletion of GSH in dopaminergic cell line, has been found to inhibit complex I activity via peroxynitrite mediated event and this could be reversed by a thiol reducing agent (Chinta and Andersen, 2006).

6.2 Thiol disulfide oxidoreductases (TDORs): Thiol disulfide oxidoreductases are a group of proteins which catalyze disulfide interchange reactions including conversion of glutathionylated proteins (PrSSG) to protein thiols (PrSH). The reduced status of cysteine sulfhydryl groups determines the native structure and biological activity of enzymes, receptors, protein transcription factors, and transport proteins in cells. TDORs help in maintaining reduced status of critical thiol groups in proteins and they comprise three major antioxidant defense systems - the thioredoxin/thioredoxin reductase system, protein disulfide isomerase and the glutathione (GSH)/glutaredoxin (Grx) system (Freedman, 1979; Holmgren, 1988). Thioredoxin and protein-disulfide isomerase have broad substrate specificity whereas, glutaredoxin specifically reduces glutathione mixed disulfides with greater efficiency than does thioredoxin (Axelsson and Mannervik, 1975; Gravina and Mieyal, 1993).

6.3 Glutaredoxin system (Thioltransferase): Glutaredoxin or thioltransferase (Grx1) is a 12-kDa cytosolic protein that has been characterized *in vitro* as a specific catalyst for the reduction of protein-glutathionyl-mixed disulfides (protein-SSG). Grx1 selectively uses GSH as reducing substrate and catalyzes thiol-disulfide interchange reactions, crucial for maintaining intracellular thiol status. Under physiological conditions the intracellular redox milieu is predominantly reducing, but pathological processes such as aging and neurodegenerative diseases shift it towards a more oxidizing state. A crucial component of the cellular redox balance is modulation of the thiol-disulfide status of critical cysteine residues on proteins and a prevalent modification of cysteine residues is reversible S-glutathionylation which protects them from further irreversible oxidation. Accumulation of protein-SSG has been reported in different cell types under a variety of oxidative conditions (Gravina and Mieyal, 1993; Jung and Thomas, 1996; Padgett and Whorton, 1998). Grx1 regulates the activity of several enzymatic proteins by modulating their thiol disulfide status by catalyzing reversible oxidoreductions using GSH (Mannervik and Axelsson, 1975, Gallogly and Mieyal, 2009). During oxidative stress, Grx1 senses redox potential and catalyzes GSH-disulfide redox reactions via two redox active cysteines. Oxidoreduction of protein disulfides is catalyzed in the dithiol mechanism wherein both active cysteine sites are required whereas reduction of glutathione mixed disulfide occurs in the monothiol mechanism. The glutaredoxin system is maintained in its active form by glutathione reductase (GR) utilizing reducing equivalent from NADPH and GSH. Mode of catalysis by Grx1 depends on the redox environment of the cell. Grx1 can transiently act as a glutathionylating enzyme in the presence of an oxidative stimulus (Equation-A) and as a deglutathionylase when oxidative stress subsides (Equation-B), thereby stabilizing proteins during oxidative stress (Axelsson and Mannervik, 1980). In first step thiolate of Grx1-N terminal cysteine attacks glutathionyl sulfur of PrSSG forming Grx-SSG intermediate and releasing reduced protein. Grx-SSG is then attacked by free GSH thus releasing reduced Grx and GSSG (Gallogly and Mieyal, 2009). Oxidized glutathione (GSSG) is restored back to the functional reduced glutathione (GSH) in the presence of glutathione reductase at the expense of NADPH (Equation-C).

$$PrSSPr' + GSH \rightarrow PrSH + GSSPr' \quad \ldots (A)$$

$$PrSSG \xrightarrow{Grx1/GSH/NADPH} PrSH + GSSG \quad \ldots (B)$$

$$GSSG \xrightarrow[NADPH]{Glutathione\ Reductase} GSH \quad \ldots (C)$$

Glutaredoxin follows a general mechanism in disulfide-catalyzed reduction, where intramolecular disulfide form of the enzyme is an intermediate in the overall reaction (Gan and Wells, 1987). The function of glutathione in the reaction is essentially to reduce the enzyme disulfide and form GSSG to complete the cycle. Glutaredoxin contains an N-terminal redox centre, the CXXC motif (Cys-Pro-Tyr-Cys) and the two-thiol groups in this motif are the source of reducing equivalents for substrate reduction. These residues being the active site of enzyme, form reversible disulfide bond during catalysis. It also contains conserved C-terminal sequences required for GSH binding (Nordstrand et al., 2000). The complete glutaredoxin and thioredoxin system is present both in cytosol and mitochondria (Gladyshev et al., 2001; Lundberg et al., 2001). Mitochondrial (Grx2) glutaredoxin shares only 34% sequence identity with Grx1, codes for an 18 kDa protein and contains CSYC as an active site motif (Johansson et al., 2004).

6.4 Role of glutaredoxin: Most important function of glutaredoxin in several organisms is maintenance of redox status during oxidative stress. Since much of the oxidative damage is attributed to functional modification of sulfhydryl protein by disulfide bond formation, glutaredoxin is thought to act as cellular repair enzyme by catalyzing reduction of disulfides. Glutathionylation is a reversible regulatory mechanism which regulates redox sensitive signal transduction pathways by either activating enzymes or inactivating them. Proteins which are known to be inactivated by glutathionylation are nuclear factor (NF-1), protein tyrosine phosphatase 1B (PTP1B-Cys215) (Barrett et al., 1999), phosphofructokinase glyceraldehyde-3-phosphate dehydrogenase (GAPDH) (Lind et al., 1998; Mohr et al., 1999), protein kinase Cα (Ward et al., 2000); nuclear factor kappa B (NFκB; (Pineda-Molina et al., 2001; Qanungo et al., 2007)); creatine kinase (Reddy et al., 2000); actin-Cys374, (Dalle-Donne et al., 2003), protein phosphatase 2A (Rao and Clayton, 2002); protein kinase A (Humphries et al., 2002); tyrosine hydroxylase (Borges et al., 2002), mitochondrial complex I (Taylor et al., 2003); IκB Kinase (IKK) (Reynaert et al., 2006). While glutathionylation inactivates the above said proteins, it is known to activate some others, such as microsomal glutathione *S*-transferase (Dafre et al., 1996), carbonic anhydrase III phosphatase-Cys186 (Cabiscol and Levine, 1996), HIV-1 protease-Cys67 (Davis et al., 1996; Davis et al., 1997), matrix metalloproteinase (Okamoto et al., 2001), hRas-Cys118, sarco/endoplasmic reticulum calcium ATPase (SERCA) (Adachi et al., 2004); and mitochondrial complex II (Chen et al., 2007).

Studies have shown that overexpression of glutaredoxin protects Akt from H_2O_2 induced oxidation and suppresses recruitment of protein phosphatase 2A to Akt resulting in sustained phosphorylation of Akt and inhibition of apoptosis and this effect can be reversed by cadmium, an inhibitor of glutaredoxin (Chrestensen et al., 2000). Besides this, glutaredoxin reduces oxidized-Akt using GSH, NADPH and glutathione-disulfide reductase thus playing a key role in protecting cells from apoptosis (Murata et al., 2003). Activity of apoptotic signal regulating kinase 1 (ASK1) is inhibited by redox regulated binding of Trx and Grx to it N and C terminal respectively (Song and Lee, 2003b). Binding of Grx to ASK1 is GSH dependent and is disrupted during oxidative conditions thus leading to the activation of apoptosis signal regulating kinase 1 (ASK-1), a mitogen-activated protein kinase kinase kinase (MAPKKK) and downstream apoptotic signaling cascade. Mitochondrial dysfunction due to complex I inhibition has been observed in animal models for PD and motor neuron disease using MPTP and L-BOAA respectively (Kenchappa et al., 2002; Kenchappa and Ravindranath, 2003). Glutaredoxin plays a critical role in the maintenance of mitochondrial complex I and its downregulation using antisense oligonucleotides further exacerbate the complex I inhibition caused by MPTP (Kenchappa and Ravindranath, 2003). Hence glutaredoxin is an important enzyme involved in redox regulation and thereby indirectly regulates redox driven signaling cascades in cells.

6.5 Thioredoxin (Trx) system: Thioredoxin system comprises of cytosolic thioredoxin (Trx1), its mitochondrial counterpart (Trx2) and thioredoxin reductase (TR). Trx1 is normally cytosolic but under stress conditions it translocates to nucleus or gets exported from the cell (Rubartelli et al., 1992). Mammalian TRs are homodimeric seleno-enzymes harboring a FAD cofactor and a penultimate COOH-terminal selenocysteine residue in their Gly-Cys-SeCys-Gly active site (Gladyshev et al., 1996). Mammalian Trx1 was first described as an electron donor for enzymes that form a disulfide during their catalytic cycle such as ribonucleotide reductase, methionine sulfoxide reductases and peroxiredoxin (Zhong et al., 2000; Rhee et al., 2005). Thioredoxin catalyzes the reduction of cysteine residues in proteins disulfides utilizing both cysteinyl residues in their (Cys-Gly-Pro-Cys) active site by forming a disulfide bond within Trx (Equation-A). The disulfide bond thus formed in Trx is further reduced by Trx reductase (TrxR; Equation-B).

$$Trx - (SH)_2 + Pr - S_2 \rightarrow Trx - S_2 + Pr - (SH)_2 \quad \ldots\ldots\ldots (A)$$

$$Trx - S_2 + NADPH + H^+ \xrightarrow{TrxR} Trx - (SH)_2 + NADP^+ \quad \ldots\ldots\ldots (B)$$

Thioredoxin system has been implicated in inflammation and induction of apoptosis. Several proteins have been identified to be redox regulated by the Trx system such as the binding of NF-κB subunit p50 to its target sequence in DNA requires the reduction of a single cysteinyl residue by Trx1 in the nucleus (Matthews et al., 1992). Trx1-catalyzed reduction also activates transcription factors including hypoxia-inducible factor1-α (Welsh et al., 2002), p53 (Ueno et al., 1999), the estrogen receptor (Hayashi et al., 1997), and c-Fos/c-Jun complexes (Hirota et al., 1997). Reduced Trx remains associated with ASK-1, thus suppressing its kinase activity, until oxidation of Trx1 causes its dissociation and activation of ASK1. Hence, redox state of Trx1 serves as a regulatory switch for the induction of apoptosis via the JNK and p38 MAPK pathways propagated downstream through ASK1 activation. Trx binding protein-2 [TBP-2, also called vitamin D3-upregulated protein 1 (VDUP1) or Trx-interacting protein (Txnip)] was identified as a Trx binding partner (Yamanaka et al., 2000). TBP-2 binds specifically to Trx present in reduced form thereby serving as a negative regulator of Trx function (Junn et al., 2000). The mitochondrial counterpart, Trx2 is found to be essential for embryonic development (Nonn et al., 2003).

6.6 Role of redox perturbation in mitochondrial dysfunction: In response to transient elevations of ROS, thiol groups of cysteine residues switch between an inactive and an active state as a mechanism in the regulation of protein function. Sulfhydryl groups upon exposure to ROS form a highly unstable product sulfenic acid (RSOH), which further reacts to produce a disulfide or a stable higher oxidation product such as a sulfinic or sulfonic acid (RSOOH or RSO_3H). Nevertheless, stable sulfenic acid derivatives have also been detected in proteins supposedly playing an important role in redox sensing (Fuangthong and Helmann, 2002; Giles and Jacob, 2002). Thus, redox based regulation of proteins is a highly dynamic post-translational regulatory mechanism of physiological and pathophysiological importance. Moreover, the detection of reactive oxygen species (ROS) and reaction nitrogen species (RNS) as signaling molecules has spurred further studies of their possible targets. While sulfenic acid can be either reduced back to the thiolate form or get modified by reactions such as S-nitrosylation (addition of NO) or S-thiolation (disulfide bridge formation with other protein thiols or with GSH; (Filomeni et al., 2003)), stable higher oxidation product such as a sulfinic acid or sulfonic acid modifications irreversibly modify the structure and function of the proteins involved. However, as per recent reports, sulfinic modification can be reversed back to thiolate form by sulfiredoxin and sestrin 2 (Lim et al., 2008) but further evidences are required to confirm the findings.

Previous studies from our lab have demonstrated the critical role of Grx1 in maintaining function of mitochondrial complex I using MPTP mouse model of PD (Kenchappa and Ravindranath, 2003), thereby indicating that maintenance of protein thiol homeostasis and hence thioltransferase are critical for preserving the functional activity of complex I. However, mechanistic role and mode of action of a cytosolic enzyme (thioltransferase; Grx1) in protecting mitochondrial complex I was unclear. The present work has tried to address this issue.

6.7 Redox-mediated signaling: Perturbation of redox homeostasis causes reversible covalent modifications in certain molecules. These reversible modifications integrate enormous biological information and fine tune the activity of target protein and have been referred to as an early event associated with the activation of cell signaling downstream. Depending on the cell type and the proteins involved, these changes caused by redox imbalance can induce either cell survival or death.

Relevance of redox perturbation in initiating a cell death cascade in PD is not very clear. The ROS production following MPTP administration causes death via mitochondrial pathway by inhibiting the mitochondrial complex I, which further propagates ROS generation. For some autosomal forms of Parkinsonism's, mitochondrial dysfunction has emerged as a common theme of pathogenesis (Ved et al., 2005), but the mechanism of ROS targeting mitochondria, and its downstream consequences are not well understood. ROS is known to activate MAPK pathway following MPTP treatment in male mice (Karunakaran et al., 2007b). Moreover, an interesting observation is that, incidences of PD have been less in women compared to men (1:1.46). This has also been reflected in animal models of PD, but the underlying neuroprotective mechanism in females have not been yet been investigated enough. Hence, there is an urgent need to explore the differential sequence of events occurring in male and female models of PD in order to understand why and how females are protected against an oxidative insult to which males are vulnerable.

6.8 ASK1-JNK-Daxx pathway: ASK1 is a redox sensitive signal transduction system, which lies upstream to JNK and p38 MAPK and responds to external and internal redox status (Ichijo et al., 1997; Saitoh et al., 1998; Nakahara et al., 1999; Liu et al., 2000). ASK1 senses the degree of stress and triggers apoptotic signaling cascade only when cells are damaged incurably by excessive and prolonged exposure to oxidative stress. Sustained activation of JNK/p38 mediated by ASK1 is required for oxidative stress induced apoptosis.

ASK1 possesses a serine/threonine kinase domain in the centre which is flanked by N- and C-terminal to which Trx and Grx are bound respectively. Trx1 inhibits the kinase activity of ASK1 by directly binding to its N-terminal. Trx1 has a redox active site in which two cysteine residues (Cys32 and Cys35) harbor the thiol groups involved in Trx-dependent reducing activity. Whereas the reduced form [Trx- $(SH)_2$] remain bound to ASK1 in healthy cells, ROS such as H_2O_2 oxidatively modifies the bound Trx (Trx-S_2) thus causing its dissociation from ASK1 (Saitoh et al., 1998; Tobiume et al., 2002). As a result, ASK1 gains its kinase activity thus inducing auto-phosphorylation of a critical threonine residue within its kinase domain (Thr838 and Thr845 of human and mouse ASK1, respectively; (Tobiume et al., 2002). Besides binding to the N-terminal to suppress the autophosphorylation, Trx also acts as a thiol reductase and reduces the cysteine residue present at 250th position (Cys250) in ASK1 and a Cys250 mutant is unable to activate the downstream JNK induced apoptosis inspite of phosphorylation of ASK1 at Thr838 (Nadeau et al., 2009).

Daxx, the death associated protein (transcriptional repressor), is a component of ASK1 signalosome (Yang et al., 1997; Chang et al., 1998). Under physiological conditions, Daxx is confined to subnuclear structures referred to as PODs (PML-oncogenic domains; (Dyck et al., 1994). During stress conditions, such as glucose deprivation, Daxx is translocated from the nucleus to the cytoplasm where it binds to ASK1, and subsequently leads to its oligomerization and activation. Phosphorylation at serine 667 of Daxx was shown to be required for relocalization of Daxx (Song and Lee, 2004). Daxx is known to interact with nuclear DJ-1, which sequesters it within the nucleus thus suppressing Daxx-ASK1 mediated apoptosis, however, DJ-1 is reportedly mutated in some pedigrees with inherited form of Parkinson's disease (Junn et al., 2005). Hence, loss of function of nuclear DJ-1 may cause Daxx dependent apoptosis.

6.9 ASK1-p38 pathway: The p38 MAP kinase functions downstream to ASK1 as has been documented in *in vitro* models of PD (Junn and Mouradian, 2001; Park et al., 2004). Activation of p38 MAPK has been linked to stress induced cell death which could be prevented by Nrf2 (Hwang and Jeong, 2008). These kinases are highly conserved through evolution and four splice variants of p38 family have been identified: p38α, p38β (Jiang et al., 1996), p38γ (ERK6, SAPK3; (Lechner et al., 1996), and p38δ (SAPK4; (Jiang et al., 1997). All p38 kinases harbor a Thr-Gly-Tyr (TGY) dual phosphorylation motif (Hanks et al., 1988) and their activation occurs in response to extracellular stimuli such as UV light, heat, osmotic shock, inflammatory cytokines (TNF-α & IL-1), and growth factors (CSF-1; (Raingeaud et al., 1995). This plethora of activators conveys the complexity of the p38 pathway which is further complicated by the fact that activation of p38α is not only stimulus dependent but is also cell type specific (Karunakaran et al., 2008).

In the programmed cell death pathway, p38 activated by ASK1 plays a role in early stages of many acute and chronic neurodegenerative diseases such as ALS (Tortarolo et al., 2003), AD (Sun et al., 2003) and transient global ischemia (Piao et al., 2002). According to recent reports activated ASK1, p38, and caspase 3 have been observed as early presymptomatic markers in ALS. The selective activation of p38 but not JNK in motor neurons is also reported to be a part of the underlying pathogenic mechanism that makes motor neurons selectively vulnerable in ALS (Raoul et al., 2002; Holasek et al., 2005). Neuroprotection has been attained by suppressing p38 activity in cerebral ischemia (Barone et al., 2001a; Barone et al., 2001b). However, the specific role of p38 in MPTP mediated neurodegeneration and the downstream sequence of events needs to be further examined. We need to ascertain if p38 is specifically activated in neurons in response to MPP^+ and whether it is differentially activated in females thus accounting for less vulnerability compared to males.

6.10 Redox modulating enzymes and regulation by estrogen: Maintenance of redox homeostasis is vital for the physiological functioning of a healthy cell. Oxidative stimulus causes perturbation in redox homeostasis which can further modulate and trigger redox dependent signaling pathways. Both survival and apoptotic pathways are activated simultaneously and a cross talk occurs among them. Fate of the cell is decided by the dominance of either of the pathways thus leading to either survival or its demise. Maintenance of redox homeostasis and its regulation in turn depends on the activity of redox regulatory enzymes. These redox regulatory enzymes are regulated by estrogen, which is known to induce protein-disulfide thiol oxidoreductases (Ejima et al., 1999). Efforts on understanding the regulatory mechanism of redox regulating enzymes, can offer insight about their potential as theraupetic targets by enhancing their activity, which might help in slowing the progression of the disease.

7.0 Estrogen mediated neuroprotection in PD: Females of several mammalian species are known to live longer than males. Same observation in humans indicates that the dissimilarity is not due to sociological differences but rather to specific biological characteristics of the two genders. The difference could be attributed to estrogen, the gonadal steroid, which has been implicated in neuroprotection besides its primary role in reproductive functions.

7.1 Epidemiology: Clinical and epidemiological studies demonstrate that the incidence and progression of PD, varies with gender. The reason for the gender related difference is unclear but accumulating evidences show that there is an association between levels of gonadal hormone, estrogen, and incidence of Parkinson's disease. Epidemiological studies indicate that incidences and prevalence of PD is more in men compared to women with an overall incidence ratio of 1.46 in males vs. females. Gender differences are also reported in response to levodopa with better response observed in women owing to its increased bioavailability (Shulman, 2007).

7.2 Estrogen as a neuroprotective molecule: Estrogen functions as a neuroprotectant in the nigrostriatal dopaminergic system and the protective mechanisms include its antioxidant property and ability to regulate the expression of Bcl-2, brain-derived neurotrophic factor (BDNF) and glial cell-derived neurotrophic factor (GDNF) (Dluzen and Horstink, 2003). It has also been suggested to be a neuroactive hormone (De Nicola et al., 2009) since it serves as a neurotrophic molecule (Singer et al., 1999) which stabilizes neuronal functions, supports neuronal viability (Yi et al., 2008) and prevents neuronal death (Yi et al., 2008). Several *in vitro* studies have demonstrated that estrogen protects cultured neurons from glutamate excitotoxicity (Hilton et al., 2006), nutritional deprivation (Zhang et al., 2009) and various other oxidative insults. *In vivo* studies have demonstrated that female rodents during their premenopausal period are less susceptible to acute insults like cerebral ischemia (Jia et al., 2009), neurotrauma and neurotoxicity.

7.3 Estrogen receptors (ERs), structure and function: Estrogen exists in three major forms in humans and rodents: the most biologically prevalent and potent estrogen 17β-estradiol (E2), less abundant estrone (E1) and estriol (E3). Most of the estrogens exert their transcriptional actions through members of the nuclear hormone receptor superfamily, estrogen receptor-α (ER-α) (Greene and Press, 1986) and estrogen receptor-β (ER-β) (Kuiper et al., 1996; Mosselman et al., 1996; Ogawa et al., 1998), which are products of two different genes located on separate chromosomes (Enmark et al., 1997). ERs have an N-terminal domain, a DNA binding domain and a ligand binding domain towards the C-terminal. Binding of estrogen to ERs promotes their dimerization and translocation to the nucleus, where they associate with specific DNA sequences called estrogen response elements (ERE) in the promoter region of target genes (Shuler et al., 1998) thus regulating transcription (Landers and Spelsberg, 1992). Under physiological conditions, ERs are intranuclear and appear to be complementary but not redundant. ERα and ERβ have antagonistic effects when they bind to ERE elements or to AP-1 or Sp-1 site, wherein ERα elicits transcriptional activation of a particular gene, ERβ represses it. Although both the receptor types share DNA binding domains, their ligand binding domains are different, hence leading to the difference in their affinities and specificities for ligands (Kuiper et al., 1997).

7.4 Mode of neuroprotective action: Neuroprotective actions of estrogen are well acknowledged since decades, however the underlying protective mechanism(s) are still unclear. Estrogen exerts its neuroprotective action via classical/genomic pathway or by a non-classical or rapid response. In the former, the steroid hormone binds to estrogen receptor - ERα/ERβ, induces a conformational change in the receptor, causing its nuclear translocation and consequent induction/repression (regulation) of gene transcription. The response is delayed since it involves genomic events involving changes in mRNA and proteins levels. Genes regulated by estrogen signaling include the nerve growth factors, such as BDNF, neurotrophin 3, their receptors TRKA-C and p75, insulin-like growth factor 1 (IGF1) and transforming growth factor α (Sasahara et al., 2007; Shalev et al., 2009). Anti-apoptotic proteins such as BCL2 and $BCLX_L$ and specific caspase inhibitors are also regulated by estrogen, since these possess an estrogen response element (ERE) binding site in their promoter. Estrogen also targets structural proteins, such as neurofilament proteins, microtubulin associated proteins, Tau, GAP43 and proteins involved in neurotransmitter metabolism, such as choline acetyltransferase and tyrosine hydroxylase. Estrogen receptors are also known to regulate enzymes involved in detoxifying oxidative stress such as quinine reductase, gamma-glutamylcysteine synthetase and glutathione S-transferases (Montano et al., 2004). Besides classical pathway, ER also acts rapidly via its interaction with intracellular signaling cascades, by associating with G-proteins, caveolins and receptor tyrosine kinases. Estrogen induces phosphorylation of ERK1/2 (members of MAPK signaling pathway), activates cyclic-AMP-responsive element binding protein (CREB), phosphorylates Akt/PKB and also regulates intracellular Ca^{2+} levels. Another non-classical model of estrogen mediated neuroprotection suggests its binding to GPR30, a membrane bound G-protein coupled receptor instead of ERα. Other than this, its phenolic structure makes it a potent antioxidant or a free radical scavenger. It effectively prevents oxidative stress induced nerve cell death by Aβ, glutamate, superoxide anions, hydrogen peroxide and lipid peroxidation in cultured neurons. However, the effective concentration at which it functions as an antioxidant is not physiologically relevant.

7.5 Role of estrogen in PD: Prevalence and incidence of PD is more in men compared to women. Evolution of symptoms in PD and responses to levodopa (L-DOPA) treatment also show sex dependent differences (Growdon et al., 1998; Zappia et al., 2005; Shulman and Bhat, 2006). Symptoms of PD and L-DOPA induced dyskinesias are shown to be modulated by estrogens (Bedard et al., 1977; Di Paolo, 1994; Giladi and Honigman, 1995). Normally, estradiol is known to modulate DA activity at various steps of transmission including DA release and metabolism by regulating pre- and post-synaptic DA receptors and DA transporter (DAT) (Di Paolo, 1994). Sex based differences are also observed in MPTP mouse model showing a greater neurotoxic effect in males than in females (Miller et al., 1998). Moreover, estrogen pre-treatment prevents MPTP-induced depletion of striatal DA in ovariectomized female mice or male mice (Dluzen et al., 1996; Miller et al., 1998; Callier et al., 2000; Grandbois et al., 2000). The effects of estrogen on dopaminergic system are either neuroprotective or symptomatic. The neuroprotective effects refer to the capacity of estrogen to prevent or modulate insults to the dopaminergic system by altering the characteristics of disease processes affecting the dopaminergic circuitry in the brain. Symptomatic effects support both suppressive and enhancing effects of estrogen, which has been documented in humans and laboratory animals. Estrogen therapy is beneficial to women with early PD prior to initiation of L-DOPA (Saunders-Pullman et al., 1999) but not at later stages of the disease (Strijks et al., 1999). An inverse correlation between factors reducing estrogen stimulation during life and PD has been observed thus supporting the hypothesis that endogenous estrogens play a role in its development (Ragonese et al., 2004). Understanding the role of estrogen in modulating the dopaminergic system will help to adapt therapies for women with Parkinson's disease (Kompoliti, 2003). Elucidating such functional role of estrogen will guide the use of postmenopausal hormonal replacement therapy in women with Parkinson's disease or for those who are genetically at risk.

7.6 Hormone replacement therapy: In spite of the benefits promised by estrogen replacement therapy, its efficacy and efficiency remains questionable at present. There is urgent need to develop novel estrogen receptor modulators that would retain the beneficial effects of estrogens in most of the organs but would not activate development of breast cancer. Considerable advancements in research in this area has resulted in emergence of selective estrogen receptor modulators (SERMs), compounds whose agonist or antagonist activities manifest in a cell-selective manner. The clinical profiles of the first generation SERMs suggest a possibility to generate compounds which only provide the beneficial effects of estrogens. Elucidation of the mechanisms responsible for tissue selectivity will enable the rational development of improved SERMs (McDonnell, 2003). Although it is hypothesized that Estrogen Replacement Therapy (ERT) may relieve the parkinsonism symptoms and may provide an opportunity to reduce the dosage of medication in women (Shulman, 2007), beneficiary effects of ERT depends on a number of factors such as dose, route of administration, its pattern and timing. Adding further to the complexity are the different sources of estrogen and variability in activation of ER subtypes which make the molecular pharmacology of ERT multifaceted. Further research is warranted in order to decipher and appreciate the beneficiary roles of ERT in postmenopausal health.

8.0 Treatment strategies for PD and milestones covered so far: Development of a neuroprotective or neurorestorative therapy which can slow down, halt or reverse the progression of PD is still an unresolved issue. Since currently available therapies are symptomatic and based on dopamine replacement, they can neither arrest nor reverse the progression of the disease resulting in a scenario where we have no cure for PD. Half a century ago, pioneering research by neuroscientists Arvid Carlsson and Oleh

Hornykiewicz revealed that parkinsonism could be reversed temporarily by pharmacologically restoring striatal dopaminergic neurotransmission and this reversal could be accomplished by dopaminergic agonists, compounds that directly stimulate postsynaptic striatal dopamine receptors (Carlsson et al., 1977). Since then, research in pursuit for disease management and cure for PD has gathered momentum. The management of Parkinson's disease can be subdivided into three categories - protective or preventive treatment, symptomatic treatment and restorative or regenerative treatment.

i. **Protective Therapy:** None of the currently existing treatments help towards slowing the progression of PD. Earlier it was believed that selective MAO-B inhibitor, selegiline, delayed the onset of levodopa responsive disability by slowing the progression of the disease (The Parkinson study group, 1993). Unfortunately, a study in United Kingdom found significantly higher mortality among patients treated with selegiline along with levodopa than among those treated with levodopa alone (Olanow et al., 1996). It is now suggested that the protective effect observed earlier, was due to the amelioration of symptoms. The development of effective protective therapy will require further advances in our understanding of the pathogenesis of the disease.

ii. **Symptomatic treatment:** Levodopa (3,4-dihydroxy-l-phenylalanine), remains the most effective treatment for PD even today, however, it is associated with several complications (Stacy and Galbreath, 2008). The early use of levodopa results in the earlier development of complications such as motor fluctuations ("wearing-off" and "on–off" phenomena) and dyskinesias. Levodopa also increases dopamine turnover, with the formation of oxygen free radicals and peroxynitrite, thereby accelerating the progression of the disease (Fahn, 1996). With time, symptoms resistant to levodopa develop, but most of the patients continue to derive a substantial benefit from levodopa over the entire course of their illness (Lang and Lozano, 1998a). For better efficiency and efficacy, levodopa is supplemented with inhibitors for AAAD, COMT and MAO-B. With all its adverse effects this is the only drug known so far to relieve the symptoms. Besides treatment, levodopa is also being used for the confirmation of PD.

iii. **Regenerative or restorative treatment:** Long-term deep brain electrical stimulation (DBS) through implanted electrodes in brain, provide remarkable benefit to people suffering from neurological motor disorders. Stimulation of ventral intermediate nucleus of the thalamus and subthalamic nucleus can substantially reduce the magnitude of symptoms in PD. Although, the mechanism through which DBS confers relief is not understood, but striking similarities in the clinical effects obtained following surgical lesions and DBS suggest that it acts via disruption of inhibition of neuronal activity similar to pallidotomy, where surgical reduction of excessive inhibitory output of the internal segment of the globus pallidus to the motor thalamus or reduction of the excessive drive of the subthalamic nucleus to both output components of the basal ganglia (internal segment of the globus pallidus and pars reticulata of the substantia nigra), increases the activation of pre-motor cortices and result in nearly normal state (Wolters, 2007).

iv. **Cell and gene therapy for PD:** Advancements in experimental field of restorative neurology with implantation of cells and transfer of genes to treat patients with neurological disorders are in progress. Stem cells as well as dissected fetal tissue containing DA neurons have been used for transplantation in PD patients (Isacson et al., 2001; Mendez et al., 2005) and evidence show that transplanted fetal DA cells in patient are not destroyed whereas degeneration of native dopaminergic neurons progresses along with the disease (Kordower et al., 1995; Piccini et al., 1999). DA neurons have also been shown to reinnervate the brain and restore DA transmission (Chung et al., 2006; Ferrari et al., 2006). Safe delivery of trophic factors and other theraupetic proteins effectively, efficiently and in a site specific manner has also been a compelling challenge. Several DA trophic factors such as GDNF have been delivered in animal models, but inspite of offering anti-parkinsonian effects in animal models they failed to improve the symptoms in PD patients (Kordower et al., 1999; Nutt et al., 2003). Success of gene therapy depends on the safety of the gene itself, the gene delivery system and on the efficacy of the transgene product in treating the disease. Both cell and gene therapy approaches, have although generated a consensus for bringing about modification in adult brain, but they are yet to address the complexities associated with such novel therapeutic tools to work successfully in clinic.

8.1 Reasons for failures and future perspectives: With no dearth of potential candidate targets and putative neuroprotective agents for theraupetic interventions in PD, most of the neuroprotective therapies have been targeted towards factors involved in pathogenesis of cell death process such as mitochondrial dysfunction, oxidative stress, excitotoxicity, inflammation, impaired clearance of misfolded proteins, and signals involved in the apoptotic cascade. Yet, it has been impossible to come up with a therapy which has disease modifying effect. Major limitation to develop a neuroprotective therapy, is uncertainity of the precise pathogenic mechanism which causes cell death. Putative targets for therapy so far have been molecules involved in cell death, which is a secondary process; the primary event which triggers cell death and should be targeted is still 'unknown'. Secondly, there is lack of a precise animal model which manifests the exact etiopathology of PD, hence there is no reliability that molecules protective in animal models will also be clinically protective in PD. Moreover, defining the exact dose concentration of a drug for clinical trial, according to its plasma levels in animal model, is too complex and may not always be efficacious. Lastly, there is no reliable way to measure and declare with conviction whether a drug is neuroprotective inspite of showing significant benefits, since the benefits may be symptomatic or have a pharmacological confounder associated with them, and may not be only due to neuroprotection.

Although presently, no agent seems to be neuroprotective or disease modifying, there is still hope in future. Insight into the pathogenesis of PD would help to come up with better targets and nodal points which are primary and trigger the pathogenesis instead of targeting the secondary responses after the initiation of the pathogenesis. Mutation studies would help to come up with better models which exhibit the pathology more closely. Better trial experiments need to be designed and should be assessed on the basis of patient's outcome and not alone on detecting neuroprotection provided by the drug.

MATERIAL & METHODS

1. Chemicals
1.1 General chemicals
1.2 Source of antibody
1.3 Source of cDNA
1.4 Instruments
1.5 Statistical analysis

2. Constructs for overexpression & knockdown of Grx1
2.1 Overexpression of Grx1 – Subcloning of Grx1 into pCMV-MCS & pcDNA 3.1
2.2 RNA interference - Annealing and cloning of shRNA to Grx1 into mU6pro vector
2.3 Techniques used for cloning and verification of constructs

3. Cell culture studies
3.1 Culturing cell lines
3.2 Treatment strategies
3.3 Ectopic gene expression in mammalian cells (neuroblastoma)
3.4 Preparation of stable cell line
3.5 Experimental readout in cell culture experiments
3.6 Studies with primary human CNS progenitor cells

4. Studies with animals
4.1 Administration of MPTP to animals
4.2 Dissection of different regions of brain and
4.3 Processing of tissue

5. Protein estimation
5.1 Assay of Complex I (NADH: ubiquinone oxido-reductase)
5.2 Assay for measuring activity of glutathione reductase
5.3 Assay for measuring activity of thioredoxin and thioredoxin reductase
5.4 Immunoblotting
5.5 Co-immunoprecipitation
5.6 Immunocytochemistry & Immunohistochemistry
5.7 Nissls staining and Stereology

1. **Chemicals:** The list of the chemicals used in the present study and their source information is given below.

1.1 General chemicals:

L-BOAA was obtained from Research Biochemicals (Natick, MA), ICI 182,780 from Tocris Cookson (Avonmouth, UK). Tetramethyl rhodamine methyl ester (TMRM), 4-acetamido-4'-maleimidylstilbene-2,2'-disulfonic acid, disodium salt (AMS), 4-acetamido-4'-((iodoacetyl)amino)stilbene-2,2'-disulfonic acid disodium salt (AIS), 2',7'-dichlorodihydrofluorescein-diacetate (H_2DCFDA), JC-1 and MitoTracker were purchased from Molecular Probes (Eugene, Oregon, USA). Carbonyl cyanide *m* chlorophenylhydrazone (CCCP) was purchased from Aldrich (Milwaukee, WI, USA). L-Buthionine-S-R-Sulfoximine (BSO) was procured from Chemical Dynamics Corporation (New Jersey, USA). *In situ* Cell Death Detection Kit, TMR red, was obtained from Roche Diagonostics GmbH (Indianapolis, IN, USA). Cell culture products and Lipofectamine™ were purshased from Gibco BRL (Invitrogen, Carlsbad, CA). TSA Indirect Tyramide Signal Amplification kit was obtained from Perkin Elmer Life Sciences, (Boston, MA, USA).

α-Lipoic acid and N-acetyl cysteine were purchased from Fluka chemicals, Switzerland. Ubiquinone 1 was obtained as a gift from Eisai pharmaceuticals (Tokyo, Japan). Polyvinylidene Fluoride (PVDF) membrane and standard protein molecular weight markers were procured from Bio-Rad Laboratories (Hercules, CA, USA). Vectastain-ABC Elite kit was purchased from Vector labs (Burlingame, CA, USA). Protein G Sepharose 4 Fast-flow was obtained from GE healthcare (Uppsala, Sweden). T4 DNA Ligase, Alkaline Phosphatase (Calf Intestine), TaKaRa La Taq and PCR reagents were purchased from Takara Bio Inc., Shiga, Japan. The TA cloning kit was purchased from Invitrogen Corporation, CA, USA. All Restriction enzymes were purchased from

New England Biolabs, Ipswich, MA, USA. Single-stranded cDNA synthesis kit and SYBR Green supermix were purchased from Applied Biosystems (Foster city, CA 94404 USA), PCR reagents were purchased from Roche Diagnostics, Roche Biochemicals, Germany. Double stranded cDNA synthesis kits, PCR reagents were purchased from Roche Biochemicals, Germany. Real-time PCR was performed using High Capacity cDNA Reverse Transcription kit and power SYBR Green PCR Master mix from Applied Biosystems (Foster city, CA 94404 USA). Agarose for electrophoresis was purchased from Amersham Biosciences, UK. p38 MAPK inihibitor, SB239063 was obtained from Calbiochem (Darmstadt, Germany).

Following chemicals and reagents were of analytical grade and were obtained from Sigma Chemical Company (St. Louis, MO). Cyclosporin A, β-Nicotinamide adenine dinucleotide reduced form disodium salt (β-NADH), Brilliant blue G-250 (Coomassie Brilliant Blue), rotenone, ammonium persulphate (APS), Sodium dodecyl sulphate (SDS), N,N,N',N'-Tetramethylethylenediamine (TEMED), acrylamide, N,N'-methylene bis-acrylamide, phenylmethyl sulfonyl fluoride (PMSF), pepstatin, leupeptin, aprotinin, antimycin, bromophenol blue, Protease inhibitor cocktail, MPTP (1-methyl 4-phenyl 1,2,3,6-tetrahydropyridine), poly-A lysine, poly-D lysine and laminin.

All nucleic acid (DNA or RNA) extraction and purification kits were obtained from Qiagen, Japan. All other chemicals used were of analytical grade and purchased locally from Qualigens, E-Merck, India.

1.2 Source of antibody:

Name	Catalogue No.	Company
Anti-Glutaredoxin I (polyclonal)	LF-PA0017	Lab Frontier (Seoul, Korea)
Anti-Glutaredoxin II (polyclonal)	LF-PA0029	Lab Frontier (Seoul, Korea)
Anti-Thioredoxin I (polyclonal)	LF-PA0002	Lab Frontier (Seoul, Korea)

Anti-Thioredoxin II (polyclonal)	LF-PA0012	Lab Frontier (Seoul, Korea)
Thioredoxin Reductase	LF-MA0015	Lab Frontier (Seoul, Korea)
VDAC	sc-8828	Santacruz Biotechnology Inc. (Santa cruz, CA)
ANT	sc-11433	Santacruz Biotechnology Inc. (Santa cruz, CA)
SOD1	sc-11407	Santacruz Biotechnology Inc. (Santa cruz, CA)
Tubulin	T4026	Sigma (St. Louis, USA)
Anti-FLAG	F1804	Sigma (St. Louis, USA)
Estrogen receptor-α	sc-7207	Santacruz Biotechnology Inc. (Santa cruz, CA)
Estrogen receptor-β	sc-8974	Santacruz Biotechnology Inc. (Santa cruz, CA)
pASK1 (Thr845, polyclonal)	#3765	Cell signaling technology
ASK1 (H-300, polyclonal)	sc-7931	Santacruz Biotechnology Inc. (Santa cruz, CA)
Phospho-p38 MAPK Thr180/Tyr182 (3D7, monoclonal)	sc-535	Santacruz Biotechnology Inc.
Phospho-p38 MAPK Thr180/Tyr182 (12F8, monoclonal)	sc-27006	Santacruz Biotechnology Inc.
p38 (C-20, polyclonal)	sc-7152	Santacruz Biotechnology Inc.
DJ-1 (C-16, polyclonal)	sc-27006	Santacruz Biotechnology Inc.
Daxx (M-112, polyclonal)	sc-7152	Santacruz Biotechnology Inc.
p53 (DO-1, monoclonal)	sc-1315	Santacruz Biotechnology Inc.
Bax (B-9, monoclonal)	sc-7480	Santacruz Biotechnology Inc.
SAPK/JNK (polyclonal)	#9252	Cell signaling technology
Tyrosine hydroxylase (polyclonal)	AB1542	Chemicon international (Temecula,CA).
Neuronal class III β-tubulin (Tuj1;monoclonal)	T8660	Covance (Berkeley, CA)
Anti-Histone H3	#05-499	Upstate (cell signalling solutions)
Anti-Dopamine Transporter (DAT) (polyclonal)	AB1591P	Chemicon-Milipore

| GFAP (GA5) | #3670 | Cell signaling technology |
| β-catenin (H-102) | sc-7199 | Santacruz Biotechnology Inc. |

All conjugated secondary antibodies were obtained from Vector Labs (Burlingame, CA).

1.3 Source of cDNA: cDNA to human cytosolic glutaredoxin (Grx1) in pBluescript II (+/-) phagemid (Stratagene), was kindly provided by Dr. John J. Mieyal, Department of Pharmacology, Case Western Reserve University School of Medicine, Cleveland Ohio, USA. Constructs over-expressing full length human wild type DJ-1 (WT), and mutants (C106A, C53A, L166P) in pcDNA 3.1 were received as a kind gift from Prof. Tak W. Mak (The Campbell Family Institute for Breast Cancer Research, ON, Canada) and Prof. David S. Park (University of Ottawa, ON, Canada).

1.4 Instruments: High speed refrigerated centrifuge (Beckman J2-21, USA) and low speed refrigerated centrifuge (Beckman Coulter 22R, USA) were used to isolate post nuclear supernatant, nuclear extract, crude brain mitochondria and post mitochondrial supernatant. Cells were homogenized and sonicated by using Ultra-Turrex T8 (Werke) and Branson Sonifier 450 (VWR) respectively. Spectrophotometric estimations were carried out using Ultraspec III spectrophotometer (Pharmacia, LKB, Sweden) or Beckman DU-64 spectrophotometer (Beckman, Inc., USA). Benchmark plus microplate spectrophotometer system, ELISA reader (Bio-Rad, Australia) was used for colourimetric assays. Real time imaging was performed with a fluorescence imaging system with a monochromatic light source (TILL Photonics, Olympus; Germany). Scanning and documentation of immunoblots was done using Syngene Gel chemiluminiscence documentation system using GeneSnap software. ECM 830 Electroporator model no. MA1 45-0052 (BTX, Massachusetts, USA) was used for preparing stable cell lines. Bio-Rad Miniprotean electrophoresis unit and blotting apparatus was used for SDS-PAGE and immunoblotting. Microscopic studies were carried out using Zeiss inverted and upright fluorescent microscope (Axiovert 135 M) and confocal microscope (Zeiss, LSM 510 Meta).

1.5 Statistical analysis: All experiments were carried out in triplicates or more. The standard curve was plotted using known concentrations of standards and concentration of the sample was calculated from the standard curve using linear regression. The test of significance was carried out using Student's 't' test or One Way Analysis of Variance (ANOVA) followed by Student-Newman-Keuls or Dunnett's test, as appropriate. The values were considered to be statistically significant from controls if $p<0.05$.

2. Cloning strategies:

2.1 Overexpression of Grx1 - Subcloning of Grx1 into pcDNA 3.1 & pCMV-MCS:

cDNA over-expressing human Grx1 (gi|4504024|ref|NM_002064.1|GLRX) was received as a kind gift from Dr. John J. Mieyal in pBluescript II as described earlier. The cDNA, approximately 350 bp was excised from pBluescript II using BamHI and HindIII restriction enzymes and cloned into pcDNA 3.1 for expression studies. The cDNA for Grx1 was also cloned into pCMV-MCS (Stratagene) that contains the β globin intron downstream to CMV promotor, thus facilitating higher expression levels. To prepare stable cell lines, the complete Grx1 overexpression cassette was excised from pCMV-MCS using Not I restriction enzyme and cloned into Not I restriction site of pcDNA 3.1, since suppression marker for gene expression in mammalian cells (neomycin resistance gene) present in pcDNA 3.1 was not available in pCMV-MCS. Similarily cDNA for mouse Grx2, previously amplified from mouse brain RNA and cloned into pBluescript II (Karunakaran et al., 2007a) was subcloned into pCMV-MCS using HindIII and BamHI restriction sites. The whole expression cassette for Grx2 was excised from pCMV-MCS using Not I and subcloned into pcDNA 3.1 for preparation of stable cell lines.

2.2 RNA interference - Annealing of shRNA to Grx1 oligonucleotides and cloning into mU6pro vector:

Principle: RNA interference is an evolutionary conserved, gene regulatory mechanism in eukaryotic organisms. The silencing is mediated through post-transcriptional gene silencing, normally triggered by dsRNA precursors which are rapidly processed into short RNA duplexes of 21 to 28 nucleotides in length, which then guide the recognition and eventually the cleavage or translational repression of complementary messenger RNAs. Hence, artificial introduction of dsRNA's has been exploited as a tool, to study the function of a gene product by silencing the gene expression both in cultured cells and animal models. Short hairpin RNA (shRNA; ~55 nt) is a sequence of RNA with a tight hairpin turn, it utilizes U6 promotor for gene silencing. Mechanistically, shRNA are processed by Dicer, an RNA-III-type endonuclease, into short dsRNA's (~21-23 nt), which subsequently unwound and get assembled into effector complexes called RNA Induced Transcriptional Silencing (RISC). RISC guides mRNA target recognition and mediates sequence specific degradation of complementary mRNA by cleaving it in the centre of complimentary region.

Procedure: shRNA oligonucleotides for Grx1 were designed using siRNA Target Designer software (version 1.51) from Promega (http://www.promega.com/siRNADesigner/program/). The coding sequence for Grx1 along with 50 nt 5' and 3' untranslated regions (UTR's) were pasted into the input box of target designer, selecting the siSTRIKE™ U6 Hairpin Cloning Systems and promega tested hairpin (AAGTTCTCT). Three sequences were selected on the basis of optimum GC content (~50 %) and BbsI and Xba I overhangs were inserted at 5' and 3' ends respectively. Scrambled sequence was similarly cloned. The complete sense shRNA oligonucleotide containing –

5'-upstream overhang (BbsI) - target (shRNA) - loop - reverse complement - 3',
was annealed with the antisense strand and cloned into mU6pro vector kindly provided by Prof. D. Turner, (Univ. of Michigan, Ann Arbor). Similarly, the scrambled sequence was also cloned into mU6pro vector. The mU6pro vector hosts mouse U6 shRNA promoter (RNA polymerase III), an SV40 late polyadenylation site with a Bbs1 and an XbaI cloning site to allow insertion of siRNA template sequences after the first nucleotide of the U6 RNA. cDNA for eGFP is present downstream to the U6 promoter, it is flanked by BbsI and Xba I and is removed while cloning the shRNA. For some experiments, silencing cassette from mU6pro vector was subcloned into pAAV-GFP (Stratagene, Texas) and used for knockdown studies.

Procedure for cloning hairpin shRNA or scrambled oligonucleotides into the mu6pro vector:

Schematic representation of cloning of shRNA oligonucleotides in mU6pro vector:

Cloning of shRNA oligonucleotides in mU6pro vector

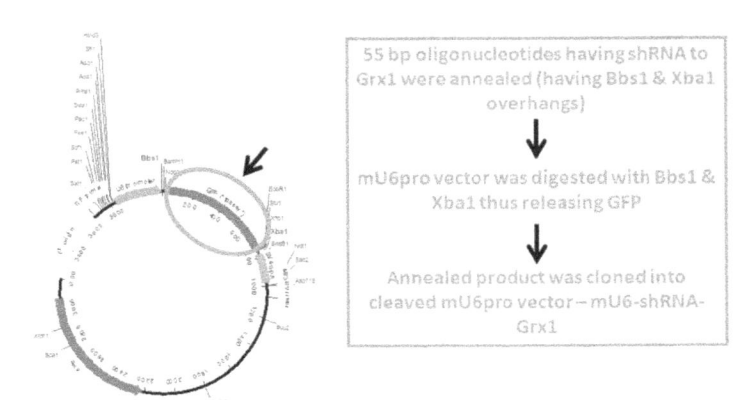

55 bp oligonucleotides having shRNA to Grx1 were annealed (having Bbs1 & Xba1 overhangs)
↓
mU6pro vector was digested with Bbs1 & Xba1 thus releasing GFP
↓
Annealed product was cloned into cleaved mU6pro vector – mU6-shRNA-Grx1

The sequence showing optimum downregulation, (60%), was used for all knockdown experiments.

The following oligonucleotide sequences (shRNA for Grx1/scrambled) were annealed and cloned into mU6pro vector:

shRNA-Grx1 -

5' - TTTGCGGATGCAGTGATCTAATAAGTTCTCTATTAGATCACTGCATCCGCTTTTT - 3'
3' - GCCTACGTCACTAGATTATTCAAGAGATAATCTAGTCACGTAGGCGAAAAAGATC - 5'

Scrambled for Grx1 –

5' - TTTGTTGGTTACGGGGTATCGATTCAAGAGATCGATACCCCGTAACCAACTTTTT - 3'
5' - CTAGAAAAAGTTGGTTACGGGGTATCGATCTCTTGAATCGATACCCCGTAACCAA - 3'

DNA oligonucleotides were resuspended in sterile miliQ at a concentration of 0.25 nmoles/ul. Annealing mix was prepared by mixing 20 µl each of oligo 1 and 2 (100 nmoles/ml final concentration) and 5 µl of 10X annealing buffer [NaCl (1M), Tris (100 mM); pH 7.4] and volume was raised to 50 µl with sterile miliQ. Annealing mix was then incubated in water heated to 95^0 C and allowed to cool at room temperature for 3 to 4 hr.

Annealing mix:

20ul oligo1 (=100nmoles/ml final concentration)
20ul oligo2 (=100nmoles/ml final concentration)
5ul water
<u>5ul 10X annealing buffer</u> (10X = 1M NaCl, 100mM Tris pH7.4)
50ul total (= 5 nmoles total of oligo)

Ligation of annealed oligonucleotides:

After cooling, the annealing mix was diluted (1:4000) in 0.5X annealing buffer and 1 µl of diluted product was ligated with 30-50 ng of Bbs1/Xba1 restriction digested and gel purified mU6pro vector in a 10 µl volume at 16°C overnight using T4 DNA ligase (Takara).

Ligation product was transformed and colonies were picked for DNA extraction according to the procedure described in "General techniques for cloning and verification of construct", later.

Colony screening and sequence verification of clones:

Colonies were picked for DNA extraction and screening by PCR amplification using either the mU6F2 or f1F as forward primers in combination with the M13R2 reverse primer (M13 reverse primer), sequence of the primer is as follows:-

M13R2 (M13 reverse 2) - 5' CACAGGAAACAGCTATGACCAT 3'
mU6F2 primer - 5' CCCACTAGTATCCGACGCCGCCATCTCTA 3'
f1F primer (f1 origin primer) - 5' CATTCAGGCTGCGCAACTGTTG 3'

An amplicon of 650 nt was observed in case of constructs having shRNA cloned whereas 1350 nt amplicon indicated recircularized plasmid.

For sequencing M13R2 (or an equivalent M13 reverse primer) was used, shRNA sequence was found 280 nt upstream to Xba I when sequenced using reverse primer. Since there is 87% homology between mouse and human Grx1, the same shRNA sequence was used to downregulate Grx1 expression in cells of both murine and human origin.

2.3 General techniques used for cloning and construct verification:

Procedure for ligation

For cloning and sub-cloning purpose, after restriction digestion and clean up, vector and insert were ligated in 1:3 ratio. 50 ng of vector was ligated with appropriate amount of insert (depending on insert size) using T4 DNA Ligase (0.5 unit; Takara) and 1µl of

10X ligation buffer in 10 µl reaction volume and the reaction mixture was incubated at 16°C overnight. Sample was stored at -20° C until transformation.

I. Procedure for transformation

For each ligation/transformation reaction, one vial of frozen TOP10/ TOP10F' competent cell was thawed on ice to which ligation mixture was added and mixed gently and incubated on ice for 30 min, followed by heat shock treatment for 60 seconds at 42° C in a water bath. 250 µl of SOC medium was added and the transformation mix was incubated in a shaking incubator set at 37° C, 225 rpm, for 45 min. The transformation mixture (100-250 µl) was plated on the LB agar plates containing the appropriate selection marker (ampicillin) and plates were incubated at 37°C for 12-14 hours. Transformed colonies were picked for plasmid isolation and screening.

II. Isolation of the plasmid

For screening of positive clones, 10 colonies were picked from LB plates and grown overnight in 7 ml of LB broth containing ampicillin. The plasmids were isolated using the Qiagen mini prep plasmid isolation kit.

Isolation of plasmids using Qiagen mini prep plasmid isolation kit

This kit was used for isolation of approximately 30 µg of high-copy plasmid DNA from 1-5 ml of overnight cultures of *E.coli* in LB medium.

Principle: Qiagen plasmid isolation protocol is based on the modified alkaline lysis procedure which releases plasmid DNA from bacteria, RNA is removed by RNase in the lysis buffer. Under low-salt and pH conditions, plasmid DNA binds to QIAGEN anion-exchange resin, where as RNA, proteins and low molecular weight impurities are removed by medium-salt wash. Finally the plasmid DNA is eluted in a high-salt buffer.

Reagents

1. Buffer P1 (Resuspension buffer): Tris-HCl (50 mM, pH 8.0) and 10 mM EDTA. RNase A was added to the above solution (100µg/ml). After addition of RNase the buffer was stored at 4°C.

2. Buffer P2 (Lysis buffer): Sodium dodecyl sulphate (SDS, 1%, w/v) in 200 mM of NaOH solution, stored at room temperature.

3. Buffer P3 (Neutralization buffer): Potassium acetate buffer (3 M, pH 5.5), stored at 4°C.

4. Buffer PE (Wash buffer): NaCl (1 M), MOPS (50 mM, pH 7.0) and 15% (v/v) isopropanol, stored at room temperature.

5. Buffer EB (Elution buffer): NaCl (1.6 M), MOPS (50 mM, pH 7.0) and 15% isopropanol, stored at room temperature.

Procedure: The bacterial cultures (1.5 ml) were harvested by centrifuging at 6000 x g for 5 min at room temperature. The supernatant was discarded and the pellet was resuspended in 250 µl of buffer P1, lysed in 250 µl of buffer P2 by gentle mixing and then incubated at room temperature for 5 min, following which 300 µl of chilled buffer P3 was added and samples were incubated in ice for 10 min. The lysate was then centrifuged for 10 min at 15000 x g and the supernatant was applied on the Qiaprep column. Columns were centrifuged for 30-60 sec at 15000 x g, flow through was discarded. Column was further washed with 0.75 ml of buffer PE by centrifuging for 30-60 sec. After discarding the flow through, it was spun for an additional 1 min to remove residual wash buffer. The Qiaprep column was then placed in a clean 1.5 ml microcentrifuge tube and 50 µl of buffer EB was added to elute the bound plasmid DNA by centrifuging the column for 1 min at 15000 x g. The eluted DNA was quantified spectrophotometrically. The isolated plasmid DNA was then checked for the presence of the insert by restriction digestion analysis and positive clones were confirmed by PCR and DNA sequencing.

III. Polymerase Chain Reaction

Principle: The Polymerase Chain Reaction (PCR) is a method to amplify DNA in vitro using a template. The principle involves enzymatic amplification of DNA fragment flanked by two oligonucleotides primers) hybridized to opposite strands of the template with the 3' ends facing each other. DNA polymerase synthesizes new DNA starting from the 3' end of each primer. Repeated cycles of heat denaturation (template), annealing of the primers and extension of the template by DNA polymerase results in amplification of the DNA fragments. The extension product of each primer serves as a template in the next cycle. Thus, the amount of the original sequence of interest is expanded in a geometric progression. This process is greatly facilitated by the use of heat stable DNA polymerase isolated from thermophilic bacteria i.e. Taq DNA polymerase which functions most effectively at relatively high temperatures.

Procedure: The PCR reaction was carried out in 50µl volume with following reactions components: 22.5 µl of sterile water, 2.0 µl of diluted template (~500 ng), 5.0 µl of 10X PCR buffer, 6.0 µl of dNTP mix (containing 10 mM of each dNTP), 8.0 µl of $MgCl_2$ (4.0 mM), 6.0 µl of forward and reverse primers (0.4µM), and 0.5 µl of (2.5 units) Takara La Taq enzyme. Thermocycling conditions used for the PCR reaction were - initial denaturation step at 94° C for 3 min followed by 40 cycles of 94° C for 1 min, 62° C for 1 min and 72°C for 2 min and a final 72°C extension for 10 min. Both negative and positive controls were included in the PCR reaction. The PCR products (20 µl) were separated by gel electrophoresis using a 1.5% agarose gel containing ethidium bromide to verify the cloning for shRNA oligonucleotides and clones showing an amplicon of 650 nt were verified by sequencing and grown for further use.

IV. DNA Sequencing

Principle: DNA sequencing is the process of determining the exact order of the bases A, T, C and G in a strand of DNA. The most commonly used method of sequencing DNA, the dideoxy or chain termination method, was developed by Fred Sanger in 1977. It is based on the use of dideoxy bases, which lack 3'-hydroxyl group required for binding with the next incoming nucleotide. Hence during DNA amplification, DNA polymerase moves along the template and continues to add bases until addition of a dideoxy base, which terminates further elongation of the chain, resulting in a set of DNA chains of different lengths. The bases are tagged with fluorescence chemical and can be read by laser thus interpreting the sequence.

Procedure: The DYEnamic ET terminator kit (MegaBACE, Amersham Biosciences, USA) was used for sequencing. It consists of: -

DYEnamic ET terminator (MegaBACE) reagent premix, control DNA: M13mp18, 0.2μg/μl, control primer: (universal cycle primer, 23-mer), 2pmol/μl, ammonium acetate (7.5 M), Loading buffer; formamide (70 %), EDTA (1 mM), deionized, distilled water.

Preparation of sequencing reaction

Preparation of DNA template: The recommended amount of DNA template (0.2-2μg; 80-800 fmol) of plasmid DNA and primers were diluted in deionized water to a total volume of 11μl. For each reaction, the following reagents were mixed: - DYEnamic ET reagent premix (8μl), primer (1μl; 5μM), DNA template (11μl) to make a reaction mix (20 μl). The reaction mix was then subjected to the following thermocycling conditions: 95 °C, 20 seconds, 50 °C, 15 seconds, 60 °C, 1 minute, (20-30 cycles).

Reaction clean up, injection and data analysis

For reaction clean up, 2µl of 7.5 M ammonium acetate was added to each of the reaction tubes followed by further addition of 2.5 volumes (55µl) of 100% ethanol to each reaction. The mix was centrifuged at 12,000 rpm for 15 minutes at 4°C, for 96 well plate precipitations, mix was centrifuged at 2,500 x g for 30 minutes. Supernatant was discarded and the pellet was washed with 100-200µl of 70% ethanol. Pellet was finally resuspended in 10µl of loading buffer and injected into the analyzer and chromatogram of the DNA sequence was analyzed using sequence analyzer (software build 94B).

3. Cell culture studies:

All neuroblastoma cell lines were procured from ATCC (American Type Culture Collection, Manassas, VA, USA). Neuroblastoma cell lines SHSY-5Y (human), and Neuro-2a (mouse) were used for in vitro studies. Some studies were carried out with primary human CNS progenitor cells after differentiating them into neurons. The purpose of using neuroblastoma cell lines was to study the role of glutaredoxin in maintenance of mitochondrial function and redox homeostasis by either overexpressing it or by silencing its expression. SHSY-5Y is a dopaminergic neuroblastoma cell line of human origin, which expresses estrogen receptor α and β, hence it was used to study the effect of estrogen on thiol-disulfide oxidoreductases. Dopaminergic neurons derived from primary human CNS progenitor cells were used to intersect the altered signaling in response to MPP^+ which is an active metabolite of MPTP.

3.1 Culturing cell lines

Reagents:

1. Dulbecco's minimum essential medium (DMEM): Powdered DMEM (Gibco-BRL, Invitrogen) was dissolved in a litre of autoclaved milliQ, 3.7 gms of sodium bicarbonate

(molecular grade from Sigma) was added and the media was filtered through 0.22 μm membrane filters.

2. Minimum essential medium (MEM): Powdered MEM (Gibco-BRL, Invitrogen) was dissolved in a litre of autoclaved milliQ, 2.2 gms of sodium bicarbonate (molecular grade from Sigma) was added and the media was filtered through 0.22 μm membrane filters.

3. Fetal Bovine Serum (FBS): FBS was heat inactivated at 55°C prior to aliquoting and storing it at −70°C.

4. Penicillin-Streptomycin solution: 10 ml of 100X stock solution of Penicillin-Streptomycin solution was added to 1 litre of DMEM complete media.

5. DMEM Complete: 1 litre of complete DMEM consisted of sterile filtered DMEM containing 10 ml of Penicillin-Streptomycin solution, 50 ml of heat inactivated FBS (10% v/v).

6. MEM Complete: 1 litre of complete MEM consisted of sterile filtered DMEM containing 10 ml of Penicillin-Streptomycin solution, 50 ml of heat inactivated FBS (10% v/v).

7. Trypsin-EDTA solution: From the 100X stock solution, 10 ml of 1X trypsin-EDTA working solution was used in the experiments.

8. Geneticin antibiotic (G-418): 1 mg/ml stock solution of G-418 was prepared in sterile filtered PBS.

Procedure: Neuro-2a and SHSY-5Y were cultured in complete DMEM and MEM medium respectively, supplemented with 10% (v/v) fetal bovine serum, 100 units/ml penicillin-G and 100 mg/ml streptomycin. Cells (1×10^6) were first revived in 90 mm culture plates containing 10 ml of complete DMEM or MEM as applicable and allowed to adhere and grow in an incubator with 5% carbondioxide maintained at 37°C. Once the culture plates were confluent, cells were washed with sterile PBS (1X solution), trypsinized with 500 µl of trypsin-EDTA (1X solution) for 3 min at 37°C and reseeded at a dilution of 1:10 in culture flasks containing 10 ml of prewarmed complete medium. The cells, which formed a monolayer were used for further treatment studies and transfection experiments.

3.2 Treatment strategies

Treatments given to SHSY-5Y to study the effects of estrogen:

Regulation of thiol-disulfide oxidoreductases by estrogen was studied in SH-SY5Y cells since the cell line expresses estrogen receptors α & β. Cells were seeded at appropriate density and differentiated with dibutyryl cyclic AMP (DBA; 1 mM) for 24 hr, followed by treatment with 17-β-Estradiol (200 nM) for another 24 hr. In some experiments, cells were exposed to ICI 182,780 (1 nM), 30 min prior to exposure to estrogen in order to confirm the specificity of estrogen in regulation of thiol-disulfide oxidoreductases. To understand the mechanism of estrogen mediated neuroprotection in response to an oxidative insult, differentiated cells pretreated with estradiol or vehicle were treated with L-BOAA (1 mM) for a further period of 24 hr and used for different experiments.

Treatments studies in SHSY-5Y and Neuro-2a to induce oxidative stress and perform rescue experiments:

For some rescue experiments, involving the measurement of MMP in SHSY-5Y or Neuro-2a cells following either L-BOAA treatment or transfections, antioxidant α-lipoic acid (100 µM) or cyclosporine A (10 µM) were added along with L-BOAA (1 mM) or 6 hr following transfection. For co-localization experiments involving co-localization with MitoTracker, cells were treated with MitoTracker Deep Red 633 (500 nM) prior to fixation with 3.7 % formaldehyde (v/v). To study the effect of general oxidative stress, for some experiments, Neuro-2a cells were treated with either BSO (100 µM) or vehicle (0.9% saline) 24 hr following seeding. Medium was replenished with fresh BSO every 24 hr and the treatment was continued for 48 hr after which cells were collected and processed for either RNA/protein extraction, ROS measurement or fixed for immunocytochemistry or TUNEL. For all imaging studies, experiments were carried out in chamber slides (Nunc, Rochester; NY) charged with poly-L lysine (100 µg/ml) and Laminin (1 µg/ml) in 1X PBS overnight at 37^0C.

Treatment with MG132 to infer role of proteasome machinery in depletion of DJ-1:
For some experiments Neuro-2a cells were transfected with either scrambled construct or shRNA to Grx1 and harvested 48 hr later. To decipher the role of the proteasome machinery in depletion of DJ-1, MG132 (10 µM) was added to cells 0 or 7 hr prior to harvesting. To examine the fate of exogenously expressed human DJ-1, Neuro-2a cells were also co-trasfected with FLAG tagged WT DJ-1 along with shRNA or scrambled construct and then treated with MG132 (10 µM) or dimethyl sulfoxide (DMSO) for 7 hr, 41 hr post transfection. Non treated cells were harvested at 41 hr post transfection.

3.3 Ectopic gene expression/knockdown in mammalian cells:

Neuroblastoma cell lines Neuro-2a or SHSY-5Y were transiently transfected with either control backbone vectors or with shRNA/scrambled sequence to Grx1 cloned in mU6pro vector (as described earlier) for knockdown studies. For experiments using overexpression constructs, cell lines were transfected with constructs overexpressing human Grx1, Grx2 wild type or mutants of DJ-1 (C106A, C53A, L166P). Transfections were done using Lipofectamine™ 2000 or Lipofectamine LTX along with PLUS reagent (Gibco BRL, Invitrogen) according to manufacturers instructions.

Transfection

Transfections for knockdown studies using shRNA to Grx1 were done in both SHSY-5Y as well as Neuro-2a, studies involving overexpression of Grx1 were done in

SHSY-5Y, whereas studies involving overexpression of DJ-1 wild type or mutants were conducted in Neuro-2a cell line.

Principle: It is a technique of introducing nucleic acids into eukaryotic cells by non viral mechanism. It can be either physical or liposome or calcium phosphate mediated delivery.

Procedure: Cells were seeded at a suitable cell density and cultured in complete medium. 24 hr following seeding complete medium was replaced by OptiMEM. Appropriate amount of DNA and lipofectamine were diluted separately in OptiMEM, and incubated for 5 min. Diluted DNA and lipofectamine were mixed properly and incubated further for 30 min. Transfection mix was added to the cells dropwise and mixed well in the OptiMEM present in the culture plate followed by incubation of culture at 37^0 C in CO_2 incubator. OptiMEM was replaced with fresh complete medium 6 hr later and the cells were cultured further for 36-48 hr. For knockdown experiments Lipofectamine LTX and PLUS reagent was used instead of the regular lipofectamine to reduce cell toxicity. While using Lipofectamine LTX, required amount of DNA was diluted in OptiMEM and mixed well as suggested by the manufacturer. Appropriate amount of PLUS reagent was added to the diluted DNA, mixed well and the mix was incubated for 5 min. Lipofectamine LTX was added further to the mix according to manufacturers instructions and mixed well by gentle pipetting and the transfection mix was incubated at room temperature for 45 min. After incubation, the mix was added dropwise to the cell culture plate seeded for transfection and cells were incubated at 37^0 C in CO_2 incubator. OptiMEM was replaced by fresh complete medium 6 hr later after washing cells briefly with 1X PBS and the culture was grown for 48 or 72 hr. For co-transfections involving overexpression of wild type or mutants of DJ-1 and knockdown of Grx1, DJ-1 overexpression and Grx1 knockdown constructs were co-transfected simultaneously and cells in control group were co-transfected with scrambled construct and respective backbone vector to ensure, that each group is transfected with two plasmids in 1:1 ratio.

3.4 Preparation of stable cell lines:

Stable cell lines were prepared by electroporating SHSY-5Y cell lines with either backbone plasmid (pcDNA 3.1) or Grx1 overexpression construct (pCMV-Grx1 cloned in pcDNA 3.1) using ECM 830 Electroporator (MA1 45-0052, BTX, Massachusetts, USA).

Schematic for preparation of stable cell lines:

Cloning of Grx1 & preparation of overexpressing stable cell lines in SHSY-5Y

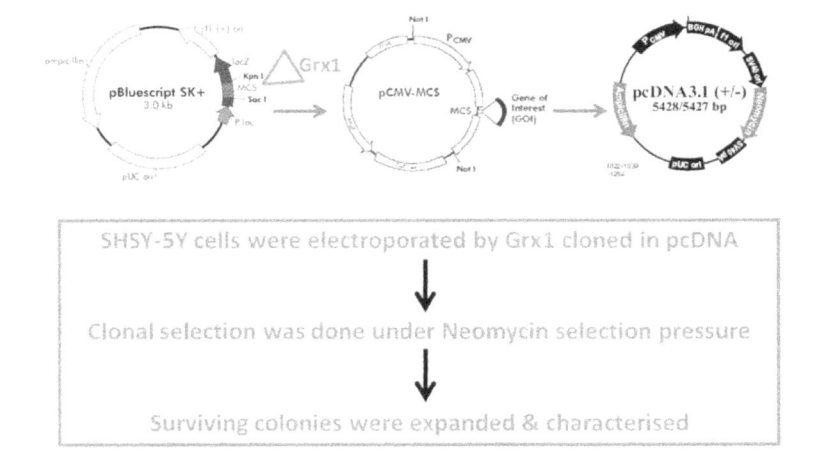

SHSY-5Y cells were electroporated by Grx1 cloned in pcDNA

↓

Clonal selection was done under Neomycin selection pressure

↓

Surviving colonies were expanded & characterised

I. Principle of Electroporation

Electroporation is the application of controlled, pulsed electric fields to biological systems. In a biological system containing a lipid bilayer, the applied pulsed electric field overcomes the field potential of the lipid bilayer, resulting in a reversible breakdown of the bilayer and a consequential formation of temporal pores in the membrane. The pores formed, are of the order of 40 to 120 nm and most of them reseal after allowing the transfer of materials into and out of the cells. During a typical electroporation process, target cells and DNA molecules are mixed together and when a pulse is delivered, the target molecules enter the cells before the pores reseal. Upon resealing of the pores, the molecules become incorporated within the cell. The eventual target site depends on the application of pulse; molecules can remain in the cytoplasm, interact with the membrane, or move into the nucleus.

II. Procedure: Electroporation using ECM 830 Electroporator was standardized to electroporate SHSY-5Y with a construct of size 7.5 kb.

Briefly, banana plugs of BTX safety stand were inserted into the HV output on front panel of ECM 830 and power switch was pressed on to initialize the electroporator. Parameter control knob was then rotated to adjust the voltage and was pushed in to select the desired voltage, in this case knob was adjusted at 200 V after selecting the 'light voltage' settings. Similarly, pulse length and number of pulses were selected using the parameter control knob, a single pulse of 1 ms was selected in this case. Further, sample, consisting of a mixture of 1×10^6 cells and 20 μg of linearized DNA digested with ScaI (pcDNA-pCMV-Grx1 and pcDNA empty vector), was prepared, mixed properly and aseptically transferred to the BTX Disposable Cuvettes Plus, cuvettes were placed in safety stand after covering the safety lid properly. The start button was then pressed to release a single pulse to the mix, after the pulse is over, the cuvette containing sample was removed from the safety stand and aseptically plated in a 100 mm culture dish containing pre-warmed medium incubated at 37^0 C in a CO_2 incubator. Following this, cells were cultured for three weeks till the colonies were formed. Each clone was picked and grown separately in a 6 well culture dish with medium containing 500 μM Geneticin for selection pressure, since this dose was found to be lethal for untransfected SHSY-5Y when standardized by gradually increasing the concentration of Geneticin. Finally each clone surviving in medium containing 500 μM Geneticin was expanded and verified by immunoblotting, quantitative real time PCR, immunocytochemistry and the clones overexpressing high levels of Grx1 were used for experiments or cryopreserved for further usage. Stable cell clones electroporated with only backbone plasmid (pcDNA) and showing Grx1 expression comparable to non electroporated cells were used as controls.

3.5 Experimental techniques used for *in vitro* (cell culture) studies:

I. Quantitative real-time PCR for assessing Grx1 expression

Verification of overexpression and knockdown for a gene was done by Quantitative Real-Time PCR, since it is a reliable and reproducible technique. Sample preparation for real time PCR involved – RNA isolation and cDNA synthesis which are briefly described as follows : -

Steps involved in quantification by real time PCR:

A. RNA isolation from cell culture:

Principle: RNA isolation from cells was done using RNeasy Protection Mini Kit (Qiagen) following the manufacturer's protocol. The RNeasy technology for RNA purification combines the selective binding properties of RNA, longer than 200 bases, to silica-based membrane which could be purified by subjecting it to microspin. Binding is facilitated by a specialized high-salt buffer system. Biological samples are lysed and homogenized in the presence of a highly denaturing guanidine-thiocyanate–containing buffer, which immediately inactivates RNases to ensure purification of intact RNA. Ethanol is then added to provide appropriate binding conditions, and the sample is then applied to an RNeasy Mini spin column, where the total RNA binds to the membrane and contaminants are efficiently washed away. High-quality RNA is then eluted in 30–100 μl water. This procedure can potentially purify all RNA molecules longer than 200 nucleotides hence it provides enrichment for mRNA since most RNAs <200 nucleotides (such as 5.8S rRNA, 5S rRNA, and tRNAs, which together comprise 15–20% of total RNA) are selectively excluded.

Procedure:

Cells were harvested, washed with 1X PBS and lysed in appropriate amount of RLT buffer (guanidine thiocyanate) containing β-mercaptoethanol (10 µl/ml) by homogenising each sample for 30 sec at 4^0 C, using a mechanical homogenizer. Equal volume of 70% ethanol was then added to the lysate and mixed well before applying the lysate to RNeasy column and centrifuged at 8000 x g at room temperature (RT) for 15 sec, flow-through was discarded. Following this, RNeasy spin column was washed with 700 µl Buffer RW1 (guanidine hydrochloride) by centrifuging for 15 s at 8000 x g at RT for 15 sec, flow-through was discarded. The column was washed twice with 500 µl Buffer RPE (contains ethanol) by centrifuging for 15 s and 2 min respectively at 8000 x g at RT, flow-through was discarded. Column was then placed in a new collection tube and centrifuged at maximum speed for 1 min to remove traces of RPE. Elution of RNA was done by placing the RNeasy spin column into a fresh 1.5 ml centrifuge tube and addition of 30-50 µl of RNase free water directly onto the spin column membrane followed by centrifuging for 1 min at 8000 x g at RT. Finally, purity and concentration of RNA was quantitated by measuring the absorbance at 260/280 nm.

B. Single stranded cDNA preparation:

Single stranded cDNA (sscDNA) was prepared by using the High-Capacity cDNA Reverse transcription Kit from Applied Biosystems (Foster city, CA 94404 USA). Reaction mix (2X) was prepared by adding following components as suggested by the manufacturer - random hexamer primers (2 µl; 10X), RNase inhibitor (1 µl), reverse transcription buffer (2 µl; 10X), MultiScribe™ reverse transcriptase (1 µl), 25X dNTP

Mix (0.8 µl; 100 mM) and 300 µl of nuclease free water. To the 2X reaction mixture 10 µl of total RNA (0.1 µg/ µl) was added and 20 µl of total reaction volume was loaded into the thermal cycler. The reaction mixture was incubated at 25^0C for 10 min followed by 37^0C for 120 min and finally reaction was stopped at 85^0C for 5 sec. The cDNA was stored at -30^0C for long-term storage and used for PCR amplification or for quantitative real-time PCR analysis.

C. **Quantitative Real-Time PCR:**

Principle: Real-time quantitative PCR allows the sensitive, specific and reproducible quantitation of nucleic acids. It is a method of simultaneous DNA quantification and amplification and is based on detection of a fluorescent signal produced proportionally during amplification of a PCR product. DNA is specifically amplified by polymerase chain reaction (PCR) and after each round of amplification, it is quantified. Methods of quantification include the use of fluorescent dyes such as SYBR green, which intercalate with double-strand DNA and modified DNA oligonucleotides (called probes) and fluoresce when hybridized with a complementary DNA.

Procedure: Quantitative Real Time PCR, being a highly reliable technique, was used as for verifying knockdown of Grx1 or overexpression of a gene at mRNA level, for each experiment. Total RNA was isolated from cells transfected with shRNA to Grx1, scrambled/empty vector and stable cell lines overexpressing Grx1. cDNA was synthesized from 1 µg of total RNA by using random hexamers. Quantitative real time PCR was performed using Power SYBR Green PCR Master Mix from Applied Biosystems according to the manufacturer's instructions to quantify the expression of mouse (Mm) and human (Hs) Grx1 using specific primers stated as -

Mm-Grx1-forward 5'-TCCTCAGTCAACTGCCTTTCA-3',

Mm-Grx1-reverse 5'-CTCCGGTGAGCTGTTGTAAA-3' and

Hs-Grx1-forward 5'-TCGATATCACAGCCACCAAAC-3',

Hs-Grx1-reverse 5'-CACTGCATCCGCCTATACAA-3'

Quantitative real time PCR was also performed to detect the mRNA levels of endogenous DJ-1 in Neuro-2a cells following Grx1 knockdown, using following primers for mouse DJ-1 –

DJ-1-forward 5'-ATGGCTTCCAAAAGAGCTCTGGT-3' and

DJ-1-reverse 5'-CCTTAGCCAGTGGGTGTGTT-3'

All reactions were carried out in triplicate, with control devoid of template. The thermocycling conditions were: 95^0 C for 10 min, followed by 40 cycles of 95^0 C for 20 sec, 60^0 C for 30 sec and 72^0 C for 40 sec for Grx1, for mouse DJ-1, annealing was done at 65^0 C instead of 60^0 C. 18S rRNA was used as an endogenous control for normalization. To verify that the used primer pair produced only a single product, a dissociation protocol was added after thermocycling, determining dissociation of the PCR products from 60°C to 95°C. Data was analyzed using the comparative threshold cycle ($\Delta\Delta Ct$) method. The results expressed as the fold difference (N) in the number of target gene copies relative to the number of 18S rRNA gene copies, were determined as follows: $N = 2\Delta\Delta Ct = 2(\Delta Ct\ target - \Delta Ct\ 18S\ rRNA)$, where $\Delta\Delta Ct$ is ΔCt target minus ΔCt 18S rRNA and ΔCt is the difference in threshold cycles for target and reference. The ΔCt values for the sample and 18S rRNA were determined by subtracting the average Ct value for the target gene from the average Ct value for the 18S rRNA gene.

II. Intracellular reactive oxygen species (ROS) measurement

Principle: The cell-permeable fluorogenic probe 2', 7'-Dichlorodihydrofluorescin diacetate (DCFH-DA) was employed to measure ROS. DCFH-DA diffuses into cells and is deacetylated by cellular esterases to non-fluorescent 2', 7'-Dichlorodihydrofluorescin

(DCFH), which is rapidly oxidized to highly fluorescent 2', 7'-Dichlorodihydrofluorescein (DCF) by ROS and this fluorescence can be measured spectrophotometrically. This fluorescence intensity is proportional to the ROS levels within the cell.

Procedure: For experiments involving knockdown of Grx1 using shRNA, ROS was measured as a read out of oxidative stress resulting from depletion of Grx1. The oxidant sensitive dye, H_2DCFDA was used to measure intracellular ROS. For qualitative determination of ROS, cells were seeded at 60% confluency 24 hr prior to transfection with shRNA to Grx1 or scambled contruct/empty vector. After 72 hr of transfection, Neuro-2a cells were loaded with 10 µM H_2DCFDA dissolved in dimethyl sulfoxide and incubated for 15 min at 37°C. Cells were then washed with PBS and imaged using appropriate filter in an inverted fluorescence microscope. For quantitative measurement of intracellular ROS, fluorimetric assay was performed. Cells were seeded at appropriate cell density, 24 hr prior to transfection with shRNA/scrambled to Grx1 construct and cultured for 72 hr or alternatively treated with BSO for 48 hr, followed by incubation of cells with H_2DCFDA for 45 min. Cells were then collected, washed with PBS and lyzed in ice cold lysis buffer (Potassium phosphate buffer containing 4 M NaCl, 0.5 M EDTA, 0.5 M EGTA, 1% Igepal, protease inhibitors), centrifuged at 10,000 rpm for 15 min at 4°C. Supernatant was diluted with PBS and used to measure emitted fluorescence at 530 nm.

III. Qualitative assessment of Mitochondrial membrane potential (MMP) using JC-1 indicator dye:

Principle: JC-1 (5,5',6,6'-tetrachloro-1,1',3,3' tetraethylbenzimidazolcarbocyanine iodide) is a cationic dye, which is sensitive to mitochondrial membrane potential (MMP) and it indicates mitochondrial depolarization by shifting its fluorescence emission from red (~590 nm) to green (~525 nm). The shift in membrane potential is observed as a decrease in ratio of red to green intensity with the loss of MMP. The dye permeabilizes within the cell and then into the mitochondria based on its cationic charge. It appears as red punctuate aggregates (J-aggregates) while the MMP is maintained and as green diffused monomers, when it is lost.

Procedure: After the desired treatment, cells were loaded with JC-1 (2 µg/ml) for 30 min at 37°C in the culture medium. The fluorescence intensity of both monomer (green) and aggregated (J-aggregates, red) molecules were captured immediately using appropriate fluorescence filters.

IV. Quantitation of mitochondrial membrane potential by live cell imaging:

Principle: Mitochondrial membrane potential was assessed using lipophilic cationic fluorescent probe, tetramethyl rhodamine methyl ester (TMRM), which is a cell-permeant, cationic, red-orange fluorescent dye and is readily sequestered by active mitochondria. Its localization and accumulation into the mitochondria is driven by their membrane potential. It appears within the mitochondria as red punctuate stain when the mitochondrial membrane potential is maintained but diffuses out into the cytosol and then outside the plasma membrane once the potential is lost. The dye is efficiently used for live cell imaging and quantitative measurement of MMP.

Procedure: Cells, SHSY-5Y and Neuro-2a, grown on coverslips were loaded with 500 nM TMRM for 30 min at 37°C in Hank's balanced salt solution. Real time imaging was performed with a fluorescence imaging system with a monochromatic light source (TILL Photonics, Germany) as described earlier (Sen et al., 2008). Dye loaded cells were maintained in a perfusion chamber (bath volume = 0.5 ml) mounted on the microscope stage. Fluorescence images were recorded every 5 sec using 520 nm excitation and 2X2 on-chip binning. Quantification of mitochondrial membrane potential was performed as described earlier (Sen et al., 2008). Regions of interest (ROI) were selected within several cells in each experiment for measuring change in mitochondrial membrane potential (MMP). To quantify the MMP, protonophore CCCP was added to depolarize the mitochondrial membrane and the difference in fluorescence intensity 10 sec before and 300 sec after CCCP addition was considered as the relative measure of MMP. Data is represented as percent relative change in TMRM fluorescence $(F/F_0)X100$, where F_0 and F are the fluorescence intensities at the beginning of recording and at given time points, respectively.

V. Cell viability assay:

Assay for cell viability was done to study the regulation of Grx1 by estrogen and its role in protecting cells following a toxic insult by L-BOAA.

Principle: MTT, (3-[4, 5-dimethylthiazol-2-yl]-2, 5-diphenyltetrazolium bromide) is reduced normally by mitochondrial dehydrogenases of metabolically active cells to insoluble purple formazan dye crystals. Detergent is then added to the wells which solubilizes the crystals and absorbance can be read using a spectrophotometer at 570 nm. The rate of tetrazolium reduction is proportional to the number of viable cells.

Procedure: Briefly, MTT (5 mg/ml in PBS) was added at a final concentration of 125 µg/ml to the cells after treatment or transfection and cells were incubated with MTT for 3 hr at 37°C, followed by solubilization in dimethyl formamide (50%; v/v) and SDS (20%; w/v), prior to measurement of absorbance at 570 nm.

V. TUNEL assay:

Principle: Cell death was assayed by performing TUNEL using *In Situ Cell Death Detection Kit-TMRM* (Roche). The assay quantifies cell death (apoptotic) by labelling and detection of DNA strand breaks in individual cells by flow cytometry or fluorescence microscopy. During apoptosis, DNAse activity generates double-stranded, low-molecular-weight DNA fragments (mono- and oligonucleosomes) as well as introduces strand breaks ("nicks") into the high-molecular-weight DNA. These processes can be identified by labeling the free 3'-OH termini with terminal transferase (TdT), which attaches labeled nucleotides to all 3´OH-ends (TUNEL reaction; TdT-mediated dUTP nick end labeling).

Procedure: Cells were treated with BSO (100 µM) for 48 hr or transfected with either scrambled or shRNA to Grx1 alone or along with DJ-1 overexpression constructs in 90 mm culture dishes and were re-seeded onto the chamber slides and 60 mm dishes 24 hr post transfections. Cells in chamber slides were fixed 72 hr post transfection and 48 hr following BSO treatment for TUNEL. Cell death was assessed using terminal deoxynucleotidyl transferase mediated dUTP-biotin nick end-labeling (TUNEL) method following manufacturer's protocol. Briefly, following fixation, cells were blocked with 4% BSA (containing 0.02% Triton X-100) for 20 min, washed thrice with 1X PBS and then incubated with TUNEL mix (Labelling & Enzyme solution in the ratio of 9:1) for 1 hr as suggested by the manufacturer. Following the incubation cells were washed thrice with 1X PBS, mounted with hard set DAPI and observed using fluorescence microscopy TUNEL positive cells and DAPI stained nuclei were counted in four independent experiments in each case. On average 300 cells were scored in several fields for each sample. Cells in 60 mm dishes were cultured for similar duration of the experiment and used for validation of knockdown or overexpression by quantitative real time PCR or immunoblot respectively.

VI.   Detection and estimation of redox state of proteins by AIS/AMS derivatization:

Principle: Redox status of proteins is modulated by thiol di-sulfide oxidoreductases and any oxidative insult or perturbation in thiol homeostasis modifies the oxidative state of thiol moieties present on cysteine residues of proteins. Redox status of a protein can be detected by haloalkyl derivatives such as 4-acetamido-4'-((iodoacetyl)amino)stilbene-2,2'-disulfonic acid disodium salt (AIS) and 4-acetamido-4'-maleimidylstilbene-2,2'-disulfonic acid, disodium salt (AMS). Both of these alkylating agents bind to the reduced and free thiol groups when incubated with protein lysate, thus increase the molecular weight of the protein roughly by 0.6 kDa per residue. This increase in molecular weight can be detected as a shift on the blot and represents the fraction of protein with reduced thiol moieties.

Procedure

Determination of redox status of Voltage Dependent Anion Channel (VDAC) and Adenine Nucleotide Translocase (ANT) in Grx1 knocked down cells: Cells were washed with ice cold PBS, harvested and treated with lysis buffer (1X PBS, 0.5% Igepal v/v). The lysates from cells transfected with either scrambled or shRNA-Grx1 were incubated (derivatized) with 4-acetamido-4'-((iodoacetyl)amino)stilbene-2,2'-disulfonic acid disodium salt (AIS; 30 mM; Invitrogen, CA) in Tris buffer (20 mM, pH 7.5), at 25°C for 12 hr and subjected to non-reducing SDS-PAGE, such that the free thiols in the reduced proteins were alkylated with AIS and thus could be visualized as a shift in the blot. The samples that were not derivatized with AIS were subjected to both non-reducing and reducing SDS-PAGE, followed by immunoblotting using antibody to VDAC or ANT.

Determination of redox state of FLAG DJ-1 in Grx1 knocked down cells: Cells were washed with ice cold PBS, harvested and treated with lysis buffer (1X PBS, 0.5% Igepal v/v; protease inhibitor cocktail; buffer was bubbled with nitrogen vapor for 5 min) for 30 min at 4° C under vacuum. The lysates from cells co-transfected with FLAG DJ-1 and scrambled/shRNA to Grx1 were incubated (derivatized) with 4-acetamido-4'-maleimidylstilbene-2,2'-disulfonic acid, disodium salt (AMS; 30 mM; Invitrogen, CA) in Tris buffer (20 mM, pH 7.5), at 25°C for 1 hr under vacuum and subjected to non-reducing SDS-PAGE, such that the free thiols in the reduced proteins were alkylated with AMS and thus could be visualized as a shifted band in the blot. The samples that were not derivatized with AIS were subjected to both non-reducing and reducing SDS-PAGE, followed by immunoblotting using antibody to FLAG.

3.6 Studies with primary culture of human CNS progenitor cells

Primary culture of human CNS progenitor cells, differentiated into neuronal lineage were obtained from Dr. P. Seth's lab Division of Molecular and Cellular Neuroscience; NBRC) for all experiments and were prepared in the following way:

Procedure: In their lab, human CNS progenitor cultures were prepared from 8- to 12-week-old embryos obtained from elective medical termination of first trimester pregnancies performed at the local hospital after informed consent as per the approved protocol of the institutional human ethics committee in compliance with the recommendations of the Indian Council of Medical Research, Government of India. Primary cultures of human CNS progenitor cells were prepared and cultured as monolayers on poly-D lysine coated chamber slides in serum-free neurobasal medium and neurons were derived from them as described previously (Mishra et al., 2008). Neuronal differentiation involved changing the growth factors in the progenitor media to brain-derived neurotrophic factor (BDNF, 10 ng/ml) and platelet-derived growth factor (PDGF)-A/B (10 ng/ml) for 3 weeks. Cells were seeded in 8 well chamber slide to a total count of 10,000 cells per well.

After obtaining the culture from Dr. P. Seth's lab, cells were characterised for their lineage before using them for the experiments. Almost 90% cells were found to be positive for neuronal marker, Tuj-1 as observed when immunostained for Tuj-1, glial fibrillary acidic protein (GFAP) and tyrosine hydroxylase (TH). Cells were treated with p38/JNK MAP kinase inhibitor (SB239063, 1 µM; SP600125, 5 µM) or p53 inhibitor, pifithrin-α (250 nM) 60 min before treatment with MPP^+ (10 µM) or vehicle for 24 h. Cells were then washed with PBS and fixed with 4% paraformaldehyde and immunostained for p53. Finally the cells were mounted in medium containing propidium iodide (PI) and visualized under fluorescence microscope. In some experiments the cell death was assessed using terminal deoxynucleotidyl transferase mediated dUTP-biotin nick end-labeling (TUNEL) method as described earlier. The cells were observed using fluorescence microscopy and both the TUNEL positive cells and DAPI stained nuclei were counted in four independent experiments in each case. On average 300 cells were scored in four different fields for each sample.

4. Studies with animals:

All animal experiments were carried out as per the institutional guidelines for the use and care of animals. All efforts were made to minimize animal suffering, to reduce the number of animals used and to utilize alternatives to *in vivo* techniques if available. Male and female C57BL6J mice (2 to 3 months, 25-30 g) were obtained from Central Animal Research Facility of National Brain Research Centre (NBRC). Animals had access to pelleted diet and water *ad libitum*.

4.1 Administration of MPTP to animals:

Animals were administered MPTP (30 mg/kg body weight in saline) subcutaneously, whereas control animals received vehicle alone. Animals were treated with a single dose of MPTP and sacrificed 12 or 24 hr later. In some experiments animals were administered MPTP daily for 8 or 14 days and sacrificed 24 hr after the last dose.

4.2 Dissection of different regions of brain

Mice were decapitated and the brain was removed and dissected at 4°C. Ventral midbrain and striatum were dissected as described earlier (Karunakaran et al., 2007b) and frozen in liquid nitrogen for immunoblotting and for measurement of enzymatic activity. Briefly, the skull plates were peeled away to visually locate the bregma. The brain was removed from the skull, and placed dorsal side up and bathed in buffer containing protease and phosphatase inhibitors. A coronal cut was made adjacent to inferior colliculi using a scalpel blade, a second cut was made just rostral to superior colliculi. Ventral midbrain was dissected by peeling off the cortex and removing the colliculi. To dissect striatum, cortex was peeled and the striatum was pinched off from between the lateral ventricles and corpus callosum, carefully avoiding the pallidum. The separated regions of the CNS were frozen immediately in liquid nitrogen. For some experiments animals were perfused transcardially with buffered paraformaldehyde (4% w/v) and the brain was dissected out and processed for immunohistochemistry.

4.3 Processing of tissue:

CNS regions were homogenized in 0.25 M sucrose (Potassium phosphate buffer; pH-7.4) containing protease inhibitors cocktail (Sigma Aldrich; as recommended) and phosphatase inhibitors like sodium orthovanadate 1 mM (Protein Tyrosine Phosphatase Inhibitor) or sodium fluoride 1 mM (serine/threonine phosphatase inhibitor) as appropriate and centrifuged at 1000 x g for 10 min to obtain post-nuclear supernatant. The post-nuclear supernatant was used for immunoblotting and for measuring the activity of thioredoxin, thioredoxin reductase and glutathione reductase. The post-nuclear supernatant was centrifuged again at 17,000 x g for 30 min to obtain the mitochondrial pellet. The pellet was resuspended in sucrose (0.25 M) and freeze-thawed for assay of complex I. Nuclear extracts were prepared as described (Korner et al., 1989). Briefly, the nuclear pellet obtained after 1000 x g was resuspended in nuclear extraction buffer [20 mM HEPES, (pH 7.9), containing $MgCl_2$ (1.5 mM), NaCl (0.84 M), EDTA (0.4 mM), dithiothreitol (0.5 mM) and protease inhibitors] and homogenized using a Dounce homogenizer. This was incubated at 4°C for 30 min and centrifuged at 14,000 x g for 15 min. The supernatant containing the nuclear extract was used for immunoblotting. Protein concentration was estimated by a dye-binding method (Bradford, 1976).

5. Assay of total protein levels:

Total protein levels were estimated for normalization among the samples before proceeding for enzyme activity assays and immunoblotting for both *in vitro* (cell culture) and *in vivo* (animal) studies. Total protein levels and their localization were also studied by performing immunostaining on cells and cryosections from tissues.

5.1 Protein estimation

Concentration of protein was estimated by the dye-binding method Bradford (Bradford, 1976).

Principle: The absorbance maxima of Coomassie Brilliant Blue G-250 in an acidic solution shifts from 465 to 595nm following complex formation with protein due to stabilization of the anionic forms of the dye by both hydrophobic and ionic interactions. The dye principally reacts with His, Lys, Tyr, Trp and Phe residues. Proteins form a blue coloured complex at low pH upon binding with Coomassie Brilliant Blue G-250. The complex shows absorbance maxima at 595 nm, which is directly proportional to the concentration of protein in the samples.

Reagents

Stock solution of Coomassie Brilliant Blue G-250: Coomassie Brilliant Blue G-250 (300 mg) was dissolved in 100 ml of ethanol and then 200 ml of orthophosphoric acid (85% solution) was added to it. The mixture was stirred well and stored at 4°C. Ready to use Bradford protein assay solution (BioRad) was used sometimes, following manufacturer's instructions.

Working solution of Coomassie Brilliant Blue G-250: The above stock was diluted seven times with MilliQ water and mixed well. The solution was kept at room temperature for 6 hr and then filtered and used for protein estimation.

Sodium hydroxide (0.1 N): Sodium hydroxide (0.1 N) was prepared in MilliQ water.

Standard BSA solution: BSA solution (1 mg/ml) was prepared by dissolving in MilliQ water, aliquoted and stored at –20°C until use.

Procedure: Cell lysate or tissue homogenate (2 µl) was mixed with 10 µl of 0.1 N sodium hydroxide, and volume of the solution was raised up to 100 µl with MiliQ in each well of a 48 well plate. Standard was prepared by adding different concentration of BSA to each well and mixed with the same volume of 0.1N sodium hydroxide. Plate was sealed and heated for 10 minutes at 70°C in a water bath and cooled to room temperature. To all wells, 0.9 ml of the diluted dye was added, mixed on a rotating shaker and incubated in dark for 10 minutes. The absorbance was read at 595 nm against the reagent blank in an Elisa plate reader. The protein concentration in the sample was estimated from the standard curve of bovine serum albumin.

5.2 Assay of Complex I (NADH: ubiquinone oxido-reductase)

Complex I activity was estimated in mitochondrial preparations as rotenone-sensitive NADH-ubiquinone oxido-reductase according to the method described by Shults (Shults et al., 1995).

Principle: Complex I catalyses the oxidation of NADH in the presence of ubiquinone (coenzyme Q) as an electron acceptor. In this process ubiquinone is reduced to dihydro ubiquinone. The rate of oxidation of NADH is monitored spectrophotometrically by measuring the decrease in absorbance at 340 nm.

$$NADH + H^+ + CoQ \longrightarrow NAD^+ + QH_2$$

Reagents

Potassium phosphate buffer 35 mM, pH 7.2): contains equimolar solutions of potassium dihydrogen phosphate and di-potassium hydrogen orthophosphate to attain a final pH of 7.2.

Assay buffer: was prepared using potassium phosphate buffer (35 mM, pH 7.2) containing EDTA (1 mM), BSA (1 mg/ml), sodium cyanide (2.65 mM), Magnesium chloride (5 mM) and Antimycin (2 μg/ml).

Ubiquinone 1 (2.5 mM): solution was prepared in distilled ethanol.

Rotenone (250 μM) solution was prepared by dissolving in distilled ethanol.

NADH (5 mM) solution was prepared fresh by dissolving in potassium phosphate buffer (35 mM, pH 7.2).

Procedure: The reaction mixture (total volume 0.5 ml) consisted of mitochondrial protein (70-80 μg), 10 μl of ubiquinone and the volume was made upto 0.48 ml with the assay buffer (-R+Q). The contents were incubated at room temperature for 2 minutes. After the preincubation, the reaction was initiated by addition of 20 μl of NADH solution. The enzyme kinetics was studied in the form of decrease in the absorbance monitored at 340 nm for 3 minutes. The non-specific activity was measured by adding the inhibitor of complex I, rotenone. The reaction mixture consisted of same amount of protein, 10 μl of ubiquinone and in addition 10 μl of rotenone and the volume was made up with assay buffer to 0.5 ml (+R+Q). The reaction was carried out as described earlier. The endogenous activity of the enzyme was also measured in the absence of ubiquinone and rotenone with the same amount of protein (-R-Q). Reactions were also carried out without ubiquinone but with rotenone (+R-Q). In both the instances (i.e. -R-Q and +R-Q) the activity was found to be negligible. The rotenone sensitive activity was calculated by subtracting the activity detected in the presence of rotenone (+R+Q) from the total activity (-R+Q). The activity was calculated from the following formula using the molar extinction coefficient of NADH, which is 6220 $M^{-1} cm^{-1}$.

$$\frac{\Delta A_{340}/ \min \times 1000}{0.00622 \times \text{Amount of protein (mg)}} = \text{nmoles of NADPH oxidized/min/mg protein.}$$

The enzyme activity was expressed as nanomoles of NADH oxidized/min/mg protein.

5.3 Assay for measuring activity of glutathione reductase:

Principle: The enzyme reduces glutathione disulfide to reduced glutathione with concomitant oxidation of NADPH to $NADP^+$. The formation of $NADP^+$ was measured by following a change in absorbance at 340 nm.

Reagents:

Potassium phosphate buffer (KPB; 0.05 M, pH 7.4): contains equimolar solutions of potassium dihydrogen phosphate and di-potassium hydrogen orthophosphate to attain a final pH of 7.4. EDTA (0.25 mM) was added to assay buffer.

GSSG solution: prepared in KPB (1.6 mM).

NADPH solution: prepared in KPB (1.2 mM).

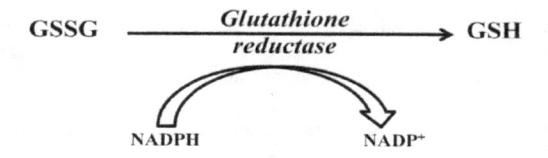

$$GSSG \xrightarrow[NADPH \quad NADP^+]{Glutathione\ reductase} GSH$$

Procedure: Reaction mixture (1.0 ml) contained GSSG (0.3 ml; 1.6 mM) and protein (20–70 µg). The volume was made up to 1.0 ml by the addition of NADPH (0.05 ml; 1.2 mM), and decrease in absorbance was monitored at 340 nm in a spectrophotometer. Blanks without protein and NADPH were run simultaneously and enzyme activity was calculated using the molar extinction coefficient of NADH, which is 6220 $M^{-1}\ cm^{-1}$.

$$\frac{\Delta A_{340}/min \times 1000}{0.00622 \times Amount\ of\ protein\ (mg)} = nmoles\ of\ NADPH\ oxidized/min/mg\ protein.$$

Enzyme activity was expressed as nmol of NADPH oxidized/min/mg protein (Horn, 1965).

5.4 Immunoblotting

Cell lysates and tissue homogenates (20 µg) were subjected to SDS-PAGE carried out as described by Lammeli and Farre (1973).

Principle: Proteins are denatured in the presence of excess of SDS and a thiol reducing agent, β-mercaptoethanol. Most of the polypeptides in this state bind to SDS in a constant weight ratio such that the charge density is identical on all polypeptides. Hence, the polypeptides migrate on the polyacrylamide gel according to their size.

Reagents

Acrylamide / bis acrylamide solution: Acrylamide (30%, w/v) and bis acrylamide (0.8%), w/v) were dissolved in MilliQ water. The solution was filtered through Whatman No 1 filter paper and stored in amber color bottle at 4°C.

4X Resolving buffer: Tris (1.5 M) was prepared in MilliQ water and the pH was adjusted to 8.8 with 4N HCL and stored at 4°C.

4X Stacking buffer: Tris (1.0 M) was prepared in MilliQ water and pH was adjusted to 6.8 with 4N HCL and stored at 4°C.

Sodium dodecyl sulphate solution (10%, w/v) was prepared in MilliQ water and stored at room temperature.

N, N, N^1, N^1-tetraethylmethylenediamine (TEMED) was obtained commercially from Sigma Chemical Company, USA.

Ammonium persulphate solution (APS, 10%, w/v) was prepared fresh in MilliQ water.

Sample buffer (2X stock): Sample buffer was prepared by mixing 1.25 ml of stacking buffer (pH 6.8), 2.5 ml of glycerol, 2.0 ml of SDS (10%, w/v), 0.5 ml of β-mercaptoethanol and 0.2 ml of Bromophenol blue (0.5%, w/v) and the total volume was made up to 10 ml with MilliQ water. The buffer was aliquoted and stored at –20°C.

Electrode buffer (10X stock) was prepared by dissolving Tris base (25 mM), glycine (192 mM) and sodium dodecyl sulphate (0.1%, w/v) in distilled water was stored at 4°C. Electrode buffer was diluted to 1X with milliQ and used for SDS-PAGE.

Procedure

Preparation of SDS resolving and stacking gels: Resolving gel of varying percentage was prepared according to the molecular weight of the target protein needed to be resolved. For each percentage of gel, the above said reagents i.e. acrylamide/ bis acrylamide, resolving gel buffer, SDS, MilliQ, APS and TEMED were added according to the specification of BioRad Gel formulations. 6% stacking gel was also prepared as suggested by BioRad.

The running gel was poured into Bio-Rad gel mould consisting of two glass plates clamped together with 1.5 mm or 1mm spacers on either side. The plates were clamped onto the stand and the running gel solution was poured slowly avoiding the air bubbles. The gel was layered with distilled water and allowed to polymerise at room temperature for 20-30 minutes. After polymerization, water was removed and stacking gel was poured and 10/15 well comb was inserted into stacking gel solution and allowed to polymerise at room temperature for 20-30 minutes. After polymerisation, the comb was gently removed.

Sample preparation: To an aliquot of the protein sample (20 μg of cell lysate or post nuclear supernatant of tissue homogenate), equal volume of 2X sample buffer was added. The sample was heated in boiling water bath for 10 minutes, cooled at room temperature and then loaded onto the gel.

Electrophoresis was carried out using constant voltage mode at 150V till the bromophenol blue dye front diffuses from the bottom of the gel. The proteins on the gel were then transferred onto PVDF membrane.

Transfer of proteins onto the PVDF and Immunological detection: The proteins were transferred from SDS-PAGE onto the PVDF membrane by the method described by Towbin (Towbin et al., 1979).

Principle: After separating the protein mixture by SDS-PAGE, they were electrophoretically transferred onto the PVDF membrane, where they bind irreversibly. The membrane was blocked to prevent non-specific binding of the antibody and then probed with the antibody of interest (primary antibody). This antigen-antibody reaction was detected by secondary antibody (anti-IgG) conjugated with alkaline phosphatase enzyme or horseradish peroxidase. The bands were visualized by incubating the membrane in the solution containing the chromogenic substrate for alkaline phosphatase, namely, nitro blue tetrazolium (NBT) and 5-bromo-4-chloro-3-indolyl phosphate (BCIP) or using chemiluminescence kit (ECL, Amersham Pharmacia Biotech, France) when the secondary antibody used was HRP labeled.

Reagents

Blotting/Transfer buffer: It consisted of Tris-HCL (25 mM), glycine (192 mM), SDS (0.1%, w/v) and methanol (20%, v/v) in distilled water.

Tris buffered saline (TBS): Tris (10 mM) and sodium chloride (0.9%, w/v) were dissolved in distilled water. The pH was adjusted to 7.5 with 1N HCL.

Blocking solution: This was prepared by dissolving skimmed milk powder (5%, w/v) in Tris buffered saline.

TBST solution: This solution was prepared by mixing Tween-20 (0.5%, v/v) in Tris buffered saline.

Primary antibody solution: Primary antibody was diluted in TBST solution.

Secondary antibody solution: Secondary antibody was diluted in TBST.

Nitro blue tetrazolium (NBT; toludine salt) solution: NBT solution (50 mg/ml) was dissolved in 70% N,N-Diethyl Formamide.

5-Bromo-4-chloro-3-indolyl phosphate (BCIP; toludine salt): This solution was prepared by dissolving BCIP (50 mg/ml) in 100% N,N-Diethyl Formamide.

Staining solution: This consisted of Tris-HCL (100 mM, pH 9.5) in distilled water containing magnesium chloride (50 mM).

Procedure: SDS-PAGE of proteins was carried out as described above. After electrophoresis, the gel was incubated in blotting buffer for 5-10 minutes and used for transfer onto PVDF membrane. Thick fiber pad of the size of the gel prewetted in the blotting buffer was placed on the transfer apparatus. Whatmann No. 2 filter paper was layered on the fiber pad after soaking it in transfer buffer avoiding trapping of air bubble between the filter paper. PVDF membrane priorly activated with methanol, was cut to the size of the gel and prewetted with blotting buffer and placed on the filter paper. The gel was then transferred onto the membrane avoiding the air bubbles. One thicker prewetted filter paper was placed gently on the gel. The cassette was closed carefully and proteins were electrophoretically transferred from gel onto the membrane at a constant current. After transfer, membrane was briefly washed with TBS and transferred into blocking solution for 2 hr at room temperature. Later the membrane was washed thrice (5 min each) in TBST with gentle shaking at room temperature and incubated with primary antibody prepared in TBST for 12 hr at 4°C or for 4 hr at room temperature with gentle shaking followed by four washes with TBST (5 min each). The membrane was then incubated with secondary antibody conjugated with alkaline phosphatase or horseradish peroxidase (HRP) as appropriate for 1 hr at room temperature with gentle shaking. After the incubation, membrane was washed five times with TBST (5 min each) followed by three washes with TBS for 5 min. Immunorecative bands were developed by adding 68 µl of NBT and 34 µl of BCIP to 10 ml of staining buffer for secondary antibody conjugated with alkaline phosphatase. After the development of colour, the reaction was stopped by washing the membrane in water, scanned and quantitated the band intensity using densitometry. For the blot incubated with secondary antibody conjugated with HRP, chemiluminescence kit (Amersham Biosciences) was used to develop the blot. Blots were normalized with β-tubulin or histone as appropriate. The relative intensities of the bands were expressed as relative optical density.

5.5 Co-immunoprecipitation

Co-immunoprecipitation was done to study the interaction between Grx1 and ectopically expressed FLAG tagged DJ-1. Antigen was pulled down using antibody to FLAG and blot was probed with antibody to Grx1 to detect their interaction. Similarly, a reverse IP was done using antibody to Grx1 and probing for FLAG-DJ-1 as an interacting partner.

Principle: Immunoprecipitation (IP) is the technique of precipitating an antigen out of solution using an antibody specific to that antigen and is used to enrich that antigen to some degree of purity. Co-immunoprecipitation involves precipitating an antigen with a specific antibody, under conditions wherein all its interacting partners remain bound to it and then probing for its interacting proteins or protein complexes. The protein complexes, once bound to the specific antibody are removed from the bulk solution by capturing it with an antibody-binding protein attached to a solid support which can be agarose / sepharose beads (slurries). Following the initial capture of a protein or protein complex, the solid support is washed several times to remove any proteins not specifically and tightly bound through the antibody. After washing, the precipitated protein(s) are eluted and analyzed using gel electrophoresis, mass spectrometry, western blotting, or other methods for identifying constituents in the complex.

Buffers and reagents

IP buffer: 50 mM Tris–HCl pH 8.0, 150 mM sodium chloride, PMSF (1 mM), 1% Igepal with a cocktail of protease inhibitors

Wash Buffer: 50 mM Tris–HCl pH 8.0

Protein G-Sepharose bead calibration: 150μl of beads (supplied in 20% ethanol) were washed with wash buffer thrice by centrifuging at 12,000 x g for 20 sec each, discarding the supernatant each time. Slurry of beads (50%) was prepared in IP buffer, mixed well and stored at 4^0 C before use.

Procedure: Cells overexpressing Grx1 and FLAG-DJ-1 were lysed in buffer containing 1X PBS, 0.5 % Igepal and protease inhibitor cocktail, 36 hr following transfection and lysate was used for co-immunoprecipitation. To the total protein concentration of 1 mg, 50 μl of Protein-A Sepharose was added and the total volume was made up to 1 ml using IP buffer. The samples were incubated at 4°C for 1hr with gentle agitation for pre-clearing to remove non-specific interactions due to endogenous IgG's. The matrix was pelleted down by centrifuging at 12,000 rpm for 20 sec (4°C) and pre-clear supernatant was collected without disturbing the pellet and transferred to a fresh eppendorf and the pellet was discarded. To the pre-cleared solution, 2 μg of Grx1/FLAG antibody was added and incubated at 4°C for 10 hr with gentle agitation. After this period, 50 μl of Protein-A Sepharose was added to each tube and incubated further for a period of 6 hours. Centrifuging the samples at 12000 x g for 5 minutes retrieved the Protein A-Sepharose-antibody-antigen complex. The supernatant was carefully aspirated and discarded. The pelleted matrix was washed 3 times with 1 ml of IP buffer followed by 2 washes with wash buffer, each time vortexing mildly and centrifuging at 12,000 x g for 1 minute, and discarding the supernatant. After the final wash the pellet was suspended in 20 μl of 5X sample buffer. The samples were then vortexed briefly, boiled for 10 minutes, spun and the supernatant was loaded onto the gel for western blotting analysis.

Immunostaining: (Immunocytochemistry/Immunohistochemistry)

Principle: The in-situ detection of protein (antigen) of interest in histological tissue preparation or sub cellular localization of the protein in individual cells is carried out using specific primary antibody, which recognizes the protein, also called as "antigen-antibody reaction". The presence of these reaction products can be shown and categorized accurately in the individual cells or tissue by the marker linked secondary antibody normally IgG), raised against the primary antibody.

I. Immunocytochemistry (in cell culture):

Immunocytochemistry is a more specific term used for immunostaining performed in cells to infer subcellular localization of the target protein. Immunocytochemistry was performed in neuroblastoma cells (SH-SY5Y and Neuro-2a) and in primary culture of human neurons differentiated into dopaminergic lineage. Cells were seeded in chamber slides and fixed with 4% paraformaldehyde (w/v) after the experiment (treatment or transfections), followed by immunostaining for the desired antigen (Grx1, Grx2, Trx1, Trx2, TR, DJ-1, ERα, ERβ, p53, pp38, Daxx). For colocalization studies involving localization of Grx1, Grx2 and DJ-1 in mitochondria, cells were loaded with Mito Tracker Deep Red 633 (500 nM) followed by fixation and immunostaining. Quantitation for subcellular localization of Daxx and DJ-1 was done by scoring an average of 400 cells in at least four independent experiments.

For ectopic gene expression or knockdown studies cells were transfected with control/scrambled vector or shRNA to Grx1 in 90 mm culture dishes, trypsinized 24 hr following transfection and re-seeded in chamber slides and 60 mm culture dishes. Cells in chamber slides were fixed 72 hrs after transfection and immunostaining was carried out for Grx1, Daxx and DJ-1. Cells, re-seeded in 60 mm culture dishes were cultured for similar time frame and used for confirming knockdown of Grx1 by quantitative real time PCR and overexpression of proteins by immunoblotting.

Staining with MitoTracker Deep Red 633

Principle: MitoTracker is a mitochondrion selective stain, which is sequestered and retained by active mitochondrion. These probes contain a mildly thiol reactive chloromethyl moiety which is responsible for its binding with mitochondria.

Procedure: Cells were loaded with MitoTracker Deep Red 633, 45 min prior to fixation. Briefly, medium was replaced with fresh medium containing MitoTracker (500 nm) after two washes with prewarmed complete medium and cells were incubated further at 37^0 C for 45 min. Following incubation, cells were washed twice with prewarmed medium and fixed with formaldehyde (3.7% v/v) at 37^0 C for 15 min. Cells were washed thrice with 1X PBS before proceeding for immunocytochemistry.

Procedure for immunocytochemistry: After the termination of treatment or transfection experiments, cells in chamber slides were washed twice with 1X PBS and fixed with paraformaldehyde (4% v/v) for 20 min at room temperature. Cells were washed thrice prior to blocking with BSA (4% containing 0.02% Triton-X 100 v/v) for 30 min at room temperature. After a mild wash with 1X PBS cells were incubated with desired primary antibody (prepared in 0.1% BSA) for 12 hr at 4^0 C, washed thrice for 5 min each and incubated with secondary antibody conjugated with Alexa Fluor 488/594 for 1 hr, again followed by washing for 3 times with 1X PBS. Cells were then mounted in mounting medium (Vectashield, Vector Labs) containing DAPI and visualized under the fluorescence microscope.

Signal amplification using Biotinyl Tyramide reagent: For some antigen having very low basal expression levels such as p53, signal was enhanced using TSA Indirect Tyramide Signal Amplification kit (Perkin Elmer Life Sciences, Boston, MA) with the slight modification in the protocol for immunocytochemistry described above. After incubation with primary antibody and subsequent washes, cells were incubated with secondary antibody conjugated with HRP (1:250) for 1 hr, washed with 1X PBS and incubated with Biotinyl Tyramide reagent (1:100; prepared in amplification buffer provided with the kit) for 10 min, washed again thrice and incubated with streptavidin fluorescence (1:250) for 30 min. Cells were washed 5 times with 1X PBS and mounted in mounting medium (Vectashield, Vector Labs) containing DAPI and visualized under the fluorescence microscope.

Immunohistochemistry - (IHC in animal tissue sections): Avidin-biotin amplification method based on the ability of avidin to bind four molecules of the biotin molecules was used for immunohistochemistry. The secondary antibody is conjugated with biotin, forms irreversible interaction with the avidin of Avidin-Biotin-peroxidase complex (ABC). A chromogen like 3, 3^1-diaminobenzidine tetrahydrochloride (DAB) or Nova red substrates for peroxidase are used to visualize the immunoreaction. The DAB or Nova red acts as an electron donor for peroxide and on oxidation forms a polymer, which precipitates at the reaction site and gives brown or red color respectively. In a separate set of experiments immunostaining was visualized using FITC labeled secondary antibody and the sections were counter-stained with DAPI.

Reagents

Paraformaldehyde (4%, w/v): Fresh solution was prepared by dissolving paraformaldehyde in phosphate buffer saline by heating at $50°$ C and then cooled at $4°$ C.

Phosphate buffered saline (PBS): Sodium phosphate (10 mM, pH 7.5) buffer containing 0.9% (w/v) sodium chloride was prepared in water, and pH was adjusted to 7.5.

Citrate Buffer/ antigen-unmasking solution: 10mM Citric Acid, pH 6.0.

Blocking reagent: 3% normal serum containing 0.1% triton X-100 and 1% BSA.

Primary antibody: The antibody was diluted in blocking reagent.

GB1 solution: This solution was prepared by dissolving sodium chloride (150 mM, w/v) and Tris base (110 mM, w/v) in distilled water and the pH was adjusted to 7.5 with HCL (1N).

Tris- buffered saline (TBS): Tris-HCL buffer (50 mM, pH 7.4) containing sodium chloride (0.9%, w/v) was prepared in water and the pH was adjusted to 7.4 with HCL (1N).

Biotinylated secondary antibody: Secondary antibody labelled with biotin molecule (4μl from Vector laboratories) was diluted to one ml in 1x PBS.

Vectastain-Elite-ABC reagent (Avidin biotin-peroxidase complex): This reagent was prepared by mixing equal volume (20 μl) of Reagent A and Reagent B in 1 ml of PBS buffer, mixed immediately and allowed to stand for about 30 min before use.

Substrate solution: Diamino benzidine (DAB, 10 mg) and 20 μl of hydrogen peroxide (30%, v/v) were dissolved in 1 ml of MilliQ.

Fluorescein conjugated secondary antibody: Fluorescein conjugated secondary antibody (1:500) was diluted in PBS buffer.

Procedure: MPTP treated or control (saline injected) C57BL6J mice were anaesthetized with ether and perfused transcardially with PBS buffer and subsequently with 4% paraformaldehyde in PBS buffer. The brain was removed, post fixed overnight and processed for immunohistochemical analysis. Cryotome sections (30 μm thick) were cut in the coronal plane. Sections were cut at the level of striatum and midbrain for MPTP treated animals.

Sections were rinsed in PBS thrice for 5 min. Endogenous peroxidase activity was blocked by incubating the sections in 3% H_2O_2 for 20 min. Sections were washed again thrice for 5 min. Antigen retrieval was done on these sections to unmask the antigen binding sites using antigen unmasking solution in a pressure cooker. Briefly, pressure cooker was pre-heated with Coplin jar containing citrate buffer until temperature reaches 95-100°C after which slides were immersed into it. Sections were incubated in the Coplin jar for 20-40 minutes and then removed and allowed to cool for 20 minutes. The sections were then washed with PBS (5 min x 3), incubated with blocking serum for 1 hr, rinsed with PBS and further incubated with antibody to Daxx (1:200), tyrosine hydroxylase (TH; 1:500), pp38 (1:100) separately. Sections for TH were incubated at 37°C overnight at room temperature; Sections for pp38 were incubated at 4°C overnight and for 48 hr for Daxx.

Specific conditions for immunohistochemistry for Daxx: The sections were washed in 1X PBS (five washes for 3 min each) followed by incubation with secondary antibody labelled with biotin (1:250 dilution) for 1:30 hr at room temperature. After further washing, they were incubated with VECTASTAIN Elite ABC reagent for 1 hr 30 min at room temperature. The sections were then washed briefly in PBS. The colour was developed using diaminobenzidine or nova red. The sections were further washed with PBS (five washes for 3 min each), dried, dehydrated with xylene twice 5 min each and mounted using DPx. The same procedure was followed for the negative control except for the primary antibody was substituted for normal rabbit IgG.

Specific conditions for immunohistochemistry for pp38 and TH: The sections were washed in 1X PBS (5 x 3 min) followed by incubation with secondary antibody labeled with HRP (1:250 dilution) for 1:30 hr at room temperature. After further washing, they were incubated with biotinylated tyramide (1:50; Perkin Elmer Life Sciences) for 15 min at room temperature. The sections were then washed briefly and incubated with streptavidin fluorescein/Texas red (1:250) for 1 hr, washed again, dried and mounted with vector shield containing DAPI. The same procedure was followed for the negative control except for the primary antibody was substituted for normal rabbit IgG.

5.6 Nissl/Thionin staining: This method is used for the detection of Nissl body in the cytoplasm of neurons on formalin-fixed, paraffin embedded tissue sections, as well as frozen sections. The Nissl body will be stained purple-blue. This stain is commonly used for identifying the basic neuronal structure in brain or spinal cord tissue.

Reagents

Sodium acetate anhydrous: 13.16g/L

Acetic acid: 6ml/L

Thionin: 1% stock

Working thionin solution

To 240 ml acetate solution, add 360 ml acetic acid and 15 ml of 1% filtered thionin (pH 5.12-5.5).

Procedure

Cryotome sections were mounted on slides and air dried. For Nissl staining sections were rinsed with milliQ for 3 to 5 min followed by staining with thionin solution for 3-5 minutes, till the cells take dark purple stain. Excess stain was removed by rinsing with milliQ. Slides were then differentiated in ethyl alcohol and dioxan in the proportion of 1:1 for 5 to 30 min and checked microscopically, dehydrated in fresh dioxan twice for 5 min each followed by rinsing with xylene twice for 5 min each and mounting with Depex Polystyrene (DPX, resinous medium).

5.10 Stereology: Male and female C57BL6J mice were administered MPTP (30 mg/kg body weight/day, s.c.) once daily for 1, 8 or 14 days whereas control animals received saline. Animals were perfused transcardially with PBS followed by paraformaldehyde (4%, w/v in PBS), 24 h after the last injection and brains were post-fixed in paraformaldehyde. Stereological analysis was carried out by cutting coronal sections (30 µm) throughout the entire midbrain from a random start point and every fifth section was processed for tyrosine hydroxylase immuno histochemistry. Sections passing through rostral, middle and caudal regions of the SN were examined. The pars compacta region was delineated for stereological counting. This delineation excluded pars reticulata (SNpr), ventral tegmental area and the retrorubral area. Persons blind to the experiment counted the number of tyrosine hydroxylase positive neurons in the substantia nigra pars compacta (SNpc).

RESULTS

I. **Role of Glutaredoxins in Neuroprotection**

1) Knockdown of cytosolic glutaredoxin 1 leads to loss of mitochondrial membrane potential: Implication in neurodegenerative diseases.

2) Loss of DJ-1 by knockdown of glutaredoxin (Grx1), triggers translocation of Daxx and ensuing cell death.

3) Glutaredoxin 2, the mitochondrial glutaredoxin offer protection against MPP^+ mediated cytotoxicity.

II. **Mechanisms of cell death in Parkinson's disease and role of estrogen**

1) MPP^+ induced activation of p38 in primary culture of human dopaminergic neurons.

2) Redox driven cell death signaling cascade activated in males are restrained in females.

I. Role of Glutaredoxins in Neuroprotection

1) Knockdown of cytosolic glutaredoxin 1 leads to loss of mitochondrial membrane potential: Implication in neurodegenerative diseases.

Glutaredoxin I, a cytosolic protein prevents mitochondrial dysfunction: Glutaredoxin 1 (also known as thioltransferase; Grx1), is a cytosolic thiol disulfide oxido-reductase which reduces glutathionylated proteins to protein thiols and helps in maintaining the redox status of proteins during oxidative stress. Downregulation of Grx1 aggravates mitochondrial dysfunction in animal models of neurodegenerative diseases, such as Parkinson's and motor neuron disease. Since, Grx1 is a cytosolic protein, its ability to influence mitochondrial function is not anticipated. So, in the present study, an *in vitro* approach was adopted to understand the role of glutaredoxin in the maintenance of mitochondrial function. Grx1 was overexpressed in SHSY-5Y (stable cell lines) and silenced in both Neuro-2a and SHSY-5Y, using shRNA to understand its role in maintaining mitochondrial function.

Sub-cellular localization of Grx1: Prior to probing into the mechanistic role of Grx1 in the maintenance of mitochondrial function, sub-cellular localization of Grx1 was verified. Grx1 is a cytosolic thiol disulfide oxido-reductase, while Grx2, another member of dithiol glutaredoxin enzyme family is localized essentially in the mitochondria and nucleus (Gladyshev et al., 2001; Lundberg et al., 2001; Karunakaran et al., 2007a). In order to confirm the localization of Grx1 in the cytoplasm, Neuro-2a cells were loaded with MitoTracker (Deep Red 633) and immunostained with antibody to Grx1 or Grx2. Grx1 did not colocalize with MitoTracker confirming its cytoplasmic localization, unlike Grx2, which completely co-localized with MitoTracker (Fig. 1).

Development and characterization of an *in vitro* model to understand the role of Grx1 in maintenance of mitochondrial function: Functional analysis of a gene product

Figure 1: Localization of Grx1 and Grx2 in cytosol and mitochondria respectively.
Cells were loaded with MitoTracker Deep Red 633 (500 nM) for 45 min before fixation and subsequent immunostaining for Grx1 or Grx2. Images were captured using LSM510 META in confocal microscope.

Grx1 (green) does not colocalize with MitoTracker Deep Red 633 (red) in Neuro-2a cells indicating its cytosolic localization, whereas mitochondrial glutaredoxin 2 (green; Grx2) colocalizes with MitoTracker Deep Red. Bar represents 10 μm.

can be performed by either overexpressing or silencing the gene. Knockdown of Grx1 using shRNA was used as the model system to look into the mechanism of Grx1 mediated maintenance of mitochondrial function. Three shRNA oligonucleotide sequences were designed and cloned into mU6pro vector as described in the methods. The one showing optimum downregulation of Grx1 (~50-60%) was chosen for further use. In order to generate cells in which Grx1 was downregulated, Neuro-2a cells were transiently transfected with the chosen shRNA to Grx1 construct and consistent knockdown of Grx1 was observed after 72 hr following transfection. For control groups cells were alternatively transfected with empty vector. Transfection efficiency was checked by transfecting Neuro-2a cells by pAAV-GFP-shRNA-Grx1 and counterstained with DAPI. Number of cells expressing GFP per total number of DAPI count depicted number of transfected cells (~80%; Figure 2A). Downregulation of Grx1 was validated by immunoblotting, quantitative real time PCR and immunostaining. While expression of Grx1 was significantly reduced by 55-60% at mRNA levels as observed by quantitative real time PCR (qRT-PCR), Grx2 levels were unaltered in these cells indicating the specificity of the shRNA (Fig. 2B). The immunoblot of Grx1 from cells in which Grx1 was knocked down showed approximately 45-50% knockdown in all the experiments (Fig. 2C). The decrease in the expression of Grx1 was also validated by immunohistochemistry (Fig. 2D).

Downregulation of Grx1 generates reactive oxygen species: Since Grx1 helps in maintaining redox homeostasis and reduced environment within the cell, its downregulation resulted in generation of reactive oxygen species as observed by H_2DCFDA (2',7'-dichlorodihydrofluorescein diacetate) staining (Fig. 2E & 3) and the ensuing oxidative stress.

Figure 2: Knockdown of cytosolic glutaredoxin in Neuro-2a cells results in generation of ROS.

Neuro-2a cells were transfected with either pAAV-GFP-shRNA-Grx1 to check the transfection efficiency or with shRNA-Grx1/mU6pro vector to characterize the *in vitro* model of Grx1 knockdown by qRT-PCR, immunoblotting or immunostaining.

(A) pAAV-GFP-shRNA-Grx1 (green) represents the transfected cells, expressing shRNA-Grx1.

(B) Quantitation of knockdown of Grx1 as assessed by qRT-PCR, no change observed in Grx2 mRNA levels. Data is represented as mean ±SEM from 3 independent experiments. Asterisks indicate values significantly different from controls (p< 0.01).

(C) Immunoblots depicting Grx1 protein levels in control (con) and knockdown using shRNA (sh) in Neuro-2a cells and densitometric quantitation of immunoblot after normalization with β-tubulin. Data is represented as mean ±SEM from 3 independent experiments. Asterisks indicate values significantly different from controls (p< 0.01).

(D) Immunostaining for Grx1 in Neuro-2a cells transfected with empty vector or shRNA to Grx1. Bar represents 100 μM.

(E) Downregulation of Grx1 using shRNA in Neuro-2a cells enhances ROS production as seen by increased H_2DCFDA staining. Bar represents 100 μm.

Grx1 downregulation results in increased levels of free intracytosolic calcium ion concentration: Silencing of Grx1 also resulted in increased free cytosolic calcium ion concentration as can be seen by Fura-2 staining (Fig. 3), indicating disruption of calcium ion homeostasis triggered by mitochondrial dysfunction.

Optimum downregulation of Grx1 leads to alteration in mitochondrial membrane potential: The effect of Grx1 downregulation on mitochondrial membrane potential (MMP) was examined qualitatively using JC-1 dye in Neuro-2a cells showing Grx1 silencing after 72 hr following transfection of cells by shRNA to Grx1. JC-1 is a cationic dye which exhibits a potential dependent accumulation in mitochondria which is indicated by a shift in the fluorescence emission. In polarized mitochondria, it accumulates in aggregated form and appears as red punctate staining whereas in cells having depolarized mitochondria, it disseminates into the cytoplasm and appears as green diffused monomeric staining. A time dependent loss of MMP was observed in Neuro-2a cells, the loss was found to be most at 72 hour post-transfection correlating with the optimum Grx1 loss (Fig. 4A, B).

Loss of MMP following Grx1 downregulation: After qualitatively assessing the loss of MMP using JC-1 dye, it was further quantified by real time imaging of live cells loaded with mitochondrial potential sensitive dye TMRM in Neuro-2a cells. Under normal conditions TMRM sequesters in the polarized mitochondria and shows punctate staining. As the mitochondrial membrane potential is lost, it diffuses out into the cytoplasm and subsequently moves out of the plasma membrane. To quantify the MMP, the fluorescence intensity was measured in several regions of interest (ROI) representing TMRM stained mitochondria in live cells. Under identical dye loading protocol the TMRM fluorescence in cells transfected with shRNA to Grx1 was significantly lower as compared to those transfected with empty vector and did not decrease further with time suggesting that the

Figure 3: Downregulation of cytosolic glutaredoxin in Neuro-2a cells also results in increased free calcium ion concentration besides increased ROS generation.

Neuro-2a cells were transfected with either shRNA-Grx1 or mU6pro vector (empty vector control) & 2 hr post transfection cells were loaded with either H_2DCFDA or Fura 2 and imaged 15 min later using appropriate filter to assess ROS generation and calcium ion concentration qualitatively. Bar represents 100 µm.

Figure 4: Loss of MMP in Neuro-2a cells correlates with optimum knockdown of Grx1. Cells were transfected with empty vector (control for 12 and 72 hr; (Fig. 4A)) or shRNA to Grx1 (Fig. 4B) and loss of MMP was monitored after 12, 24, 48 and 72 hr of transfection using JC-1 (2 µg/ml) dye. Loss of MMP in response to downregulation of Grx1 was found to be maximum after 72 hr of transfection. Green staining represents JC-1 monomer in cells with loss of MMP whereas red staining represents JC-1 aggregates in cells with intact MMP. Bar represents 120 µm.

loss of MMP has already occurred (Fig. 5B,E). To confirm that the decrease in TMRM fluorescence is indeed due to loss of MMP, the protonophore carbonyl cyanide m-chlorophenyl hydrazone (CCCP) was added to depolarize the mitochondrial membrane. CCCP induced an abrupt decrease in TMRM fluorescence reflecting the loss of MMP. The decrease in relative change in fluorescence intensity (ΔF/F) at 300 sec after CCCP addition was considered as the relative measure of MMP. Downregulation of Grx1 resulted in loss of MMP in Neuro-2a cells (Fig. 5B,E) as compared to control cells transfected with empty vector (Fig. 5A,E) and this loss of MMP was abolished by α-lipoic acid (Fig. 5C,E) or cyclosporine A (Fig. 5D,E).

Perturbation of mitochondrial membrane potential in response to Grx1 downregulation in SH-SY5Y, human neuroblastoma cell line: Effects of downregulation of Grx1 were also examined in a neuroblastoma cell line of human origin, namely the SH-SY5Y cells, which express estrogen receptors α and β and are sensitive to excitotoxicity. Following Grx1 knockdown using shRNA, decrease in levels of Grx1 protein was examined by immunocytochemistry (Fig. 6A), qRT-PCR (Fig. 6B) and immunoblot analysis (Fig. 6C). Downregulation of Grx1 in SH-SY5Y cells resulted in a similar loss of MMP as observed in Neuro-2a cells earlier. Prior exposure to SHSY-5Y with α-lipoic acid (100 µM), prevented the loss of MMP caused by Grx1 knockdown. Cyclosporine A (10 µM), an inhibitor of mitochondrial permeability transition pore (mPTP), also helped in maintenance of MMP (Fig. 6D) as examined qualitatively using the JC-1 dye.

Figure 5: Quantitative determination of loss of MMP in Neuro-2a cells in response to Grx1 knockdown.

Grx1 was downregulated in Neuro-2a cells using shRNA to Grx1. Cells were treated with vehicle after transfecting them with empty vector, or with vehicle, α-lipoic acid (100 µM) or cyclosporine A (10 µM), 6 hr after the transfection with shRNA to Grx1. MMP was measured using TMRM as the indicator dye 72 hr after the transfection. Cells were loaded with TMRM and imaged to measure change in TMRM intensity for 300 sec prior to the addition of CCCP. Loss of TMRM intensity was further measured for 300 sec after the addition of CCCP.

Cells transfected with empty vector show abrupt decrease in TMRM intensity after CCCP treatment representing sudden loss of MMP (A). Gradual decrease in TMRM intensity in cells transfected with shRNA to Grx1 represents steady loss of MMP even before CCCP treatment which further decays gradually on its addition (B). MMP was maintained in shRNA transfected cells pretreated with α-lipoic acid (C) or cyclosporine A (D). The difference in TMRM fluorescence 2 sec prior and 300 sec after CCCP addition was considered as relative measure of MMP in different groups (E). The data shown are mean ± SEM for 25 to 30 cells from 3 independent experiments in each group. Asterisk indicates values significantly different from controls ($p < 0.05$). Loss of MMP due to the Grx1 knockdown is maintained by pretreating the cells with α-lipoic acid and cyclosporine A (E). Arrow represents time point of addition of CCCP.

Loss of MMP caused by Grx1 knockdown is rescued by an antioxidant or mPTP blocker: Loss of MMP was quantified using TMRM in SH-SY5Y cells. Pseudocolor fluorescence images of TMRM loaded SH-SY5Y cells captured before and after CCCP addition revealed the MMP status in mitochondria of the cells transfected with empty vector or shRNA to Grx1 (Fig. 7A). MMP loss was measured as decrease in TMRM intensity. While there was a sharp drop in the fluorescence in cells transfected with empty vector (Fig. 7B; a,e), similar decrease was not observed in cells transfected with shRNA to Grx1 (Fig. 7B; b,e), suggesting that the MMP was already lost in these cells due to decreased levels of Grx1. However, pretreatment of cells with α-lipoic acid, a potent antioxidant (Fig. 7B; c, e) or cyclosporine A, a mitochondrial permeability transition pore blocker, (Fig. 7B; d,e) prevented the loss of MMP.

Cytosolic Grx1 protects against L-BOAA mediated MMP loss and cell death: A stable cell line overexpressing Grx1 was generated by electroporating SH-SY5Y with 20 μg of linearized construct overexpressing Grx1, followed by clonal selection. Overexpression of Grx1 was validated by immunocytochemistry (Fig. 8A), qRT-PCR analysis (Fig. 8B) and immunoblot analysis (Fig. 8C). L-BOAA, an AMPA (α-amino-3-hydroxyl-5-methyl-4-isoxazole-propionate) agonist and a neurotoxin which causes mitochondrial dysfunction due to excitotoxicity, hence it was used as a control model for Grx1 knockdown, to perturb mitochondrial function. Cells overexpressing Grx1 were subjected to L-BOAA treatment and assayed for cell death and estimation of MMP. Overexpression of Grx1 prevented the loss of cell viability observed after L-BOAA (1mM) treatment indicating that Grx1 offered protection against L-BOAA mediated cytotoxicity and consequent cell death (Fig. 8D). Exposure of SH-SY5Y cells to L-BOAA (1mM) also resulted in loss of MMP which was not seen in cells overexpressing Grx1 as observed using JC-1 dye (Fig. 8E).

Figure 6: Grx1 silencing causes MMP loss that can be prevented by thiol antioxidants and cyclosporine A.

SH-SY5Y cells were transfected with empty vector or shRNA to Grx1 and were used for quantitation of Grx1 at mRNA and protein levels and for the qualitative observation of MMP loss, 72 hr after the transfection.

(A) Immunocytochemistry for Grx1 in cells transfected with shRNA to Grx1 show decreased Grx1 levels as compared to the mock control. Bar represents 100 µm.

(B) Quantitation of knockdown of Grx1 mRNA levels as assessed by qRT-PCR in SH-SY5Y cells transfected with empty vector or shRNA to Grx1.

(C) Immunoblots showing optimal knockdown of Grx1 protein levels in cells transfected with shRNA to Grx1 and densitometric quantitation of immunoblot after normalization with β-tubulin.

(D) Loss of MMP was observed in cells transfected with shRNA to Grx1 but not in the cells transfected with empty vector. Pretreatment of cells with cyclosporine A (10 µM) or α-lipoic acid (100 µM) prevented the loss of MMP and only the red aggregates of JC-1 representing healthy cells were observed. Bar represents 120 µm.

Figure 7: Quantitative determination of loss of MMP in SH-SY5Y cells in response to Grx1 knockdown.

Grx1 was downregulated in SH-SY5Y cells using shRNA. Cells were treated with vehicle after transfecting them with empty vector, or with vehicle, α-lipoic acid (100 μM) or cyclosporine A (10 μM), 6 hr post transfection with shRNA to Grx1 and MMP was measured using TMRM as the indicator dye. Cells were loaded with TMRM and imaged to measure change in TMRM intensity for 300 sec prior to the addition of CCCP. Loss of TMRM intensity was further measured for 300 sec after the addition of CCCP.

(A) Representative pseudocolor images of cells transfected with empty vector or shRNA and treated with either vehicle, α-lipoic acid or cyclosporine A, 100 sec before and 300 sec after addition of CCCP. Bar represents 120 μm. The fluorescence profile in the cell is represented in the pseudocolor bar.

(B) Quantification of change in MMP ($\Delta\psi_m$). Cells transfected with empty vector show abrupt decrease in TMRM intensity after CCCP treatment representing sudden loss of MMP (a). Gradual decrease in TMRM intensity in cells transfected with shRNA to Grx1 represents steady loss of MMP before CCCP treatment, which further decays on adding CCCP (b). MMP was maintained in shRNA transfected cells pretreated with α-lipoic acid (c) and with cyclosporine A (d). The difference in TMRM fluorescence 2 sec prior and 300 sec after CCCP addition was considered as relative measure of MMP in different groups (e). The data shown are mean ±SEM for 25 to 30 cells from 3 independent experiments in each group. Asterisks indicate values significantly different from controls ($p < 0.05$). Loss of MMP caused by Grx1 knockdown is maintained by pretreating the cells with α-lipoic acid (c,e) and cyclosporine A (d,e). Arrow represents time point of addition of CCCP.

Figure 8: Overexpression of Grx1 prevents loss of MMP and L-BOAA mediated cell toxicity.

Grx1 overexpressing SH-SY5Y clonal cell line and control cell lines electroporated with mock empty vector were characterized for the expression of Grx1 by immunostaining (A), quantitation of mRNA by qRT-PCR (B) and immunoblot (C). SH-SY5Y cells stably overexpressing Grx1 and those electroporated with empty vector (control) were exposed to L-BOAA (1 mM) for 24 hr before determining the cell viability. Cells overexpressing Grx1 show more viability after L-BOAA exposure as compared to the control cells (D). Data is represented as mean $\pm$ SD from 3 independent experiments. Asterisks indicate values significantly different from controls ($p < 0.05$). SH-SY5Y clonal lines overexpressing Grx1 and mock control cells were subjected to L-BOAA (1 mM) treatment for 24 hr before qualitative determination of their MMP status using JC-1 (2 µg/ml). The former (Grx1 overexpressing cell line) maintained MMP following exposure to L-BOAA while control cells showed loss of MMP detected as green JC-1 monomers (E). Bar represents 120 µm.

Loss of MMP due to Grx1 knockdown or L-BOAA cytotoxicity is caused by opening of mitochondrial permeability transition pore (mPTP): Mitochondrial permeability transition pore is a large protein complex consisting of the voltage-dependent anion channel (VDAC), adenine nucleotide translocase (ANT) and cyclophillin D. Opening of mPTP always correlates with loss of MMP, however loss of MMP may not always indicate its opening. Both Grx1 knockdown and L-BOAA treatment resulted in loss of MMP as observed qualitatively by JC-1 staining and estimated by live imaging using TMRM. To check whether this loss of MMP was accompanied by subsequent opening of mPTP, SHSY-5Y cells were exposed to cyclosporine A (10 µM), an mPTP blocker prior to Grx1 downregulation or L-BOAA treatment. Loss of MMP caused by L-BOAA toxicity (Fig. 9A) and Grx1 knockdown (Fig. 9B) was prevented by pretreating the cells with cyclosporine A indicating that the observed cell loss was accompanied by mPTP opening (Fig. 9).

Role of estrogen in regulation of thiol di-sulfide oxidoreductases: SHSY-5Y cells are known to express estrogen receptors α and β and to study estrogen mediated regulation of thiol disulfide oxidoreductases, their basal expresion were examined by immunocytochemistry (Fig. 10). To examine if expression of thiol disulfide oxidoreductases such as Grx2, Trx1, Trx2 and TR is regulated by estrogen, SHSY-5Y cells were differentiated with dibutyryl cyclic AMP (DBA; 1 mM) for 24 hr and treated with 17-β estradiol (200 nM) for 24 or 48 hr further, fixed and immunostained for Grx2, Trx1, Trx2 and TR (see p.234 for effect of estrogen on GR). For some experiments cells were also treated with ICI 182,780 (1 nM) 30 min prior to exposing cells with estrogen. 17-β estradiol treatment to cells resulted in upregulation of (thioredoxin) Trx1, Trx2 (Fig. 11A, B) and (Thioredoxin reductase) TR (Fig. 12) although, levels of Grx2 were not affected by estrogen (Fig. 13). This upregulation was prevented by pre-treating the cells with ICI 182,780 suggesting the role of estrogen receptors in the regulation of these proteins.

Estrogen upregulates Grx1 and confers protection against L-BOAA toxicity: Grx1 is constitutively expressed in greater amounts in CNS of female mice and ovariectomy downregulates Grx1 rendering them more susceptible to L-BOAA toxicity (Diwakar et al., 2007). Estrogen receptors α and β are expressed in SH-SY5Y cells (Fig. 14A) and treatment with 17-β estradiol upregulated Grx1 as seen by immunostaining (Fig. 14B) and immunoblot (Fig. 14C). To determine if estrogen protects against L-BOAA mediated mitochondrial toxicity, cells were treated the with 17-β estradiol (200 nM) 24 hr prior to treatment with L-BOAA (1 mM). MMP was monitored qualitatively using JC-1 dye. Loss of MMP caused by L-BOAA was prevented in the cells pretreated with 17-β estradiol, while cells treated with vehicle were not protected (Fig. 14D). To further confirm the neuroprotection mediated by estrogen the cell viability was examined. Cells were treated with vehicle or 17-β estradiol (200 nM) for 24 hr prior to the treatment with L-BOAA (500μM) for further period of 24 hr and cell viability assessed. As compared to the vehicle treated cells, those exposed to 17-β estradiol had greater viability indicating that estrogen pretreatment provided protection against L-BOAA mediated toxicity (Fig. 14E).

Quantitation of loss of MMP using TMRM revealed that cells pretreated with 17-β estradiol maintained MMP (Fig. 15C) like controls (Fig. 15A), as compared to the vehicle treated cells, following exposure to L-BOAA (Fig. 15B). The neuroprotection offered by 17-β estradiol was similar to that seen in cells overexpressing Grx1 (Fig. 15D). Therefore, both overexpression of Grx1 or pretreatment of cells with estrogen, which in turn upregulates Grx1, provide protection against L-BOAA mediated toxicity in SH-SY5Y cells (Fig. 15E).

Figure 9: Loss of MMP caused by Grx1 knockdown and L-BOAA toxicity was accompanied by mPTP opening.

SHSY-5Y cell were exposed to vehicle/cyclosporine A (10 µM) 60 min prior to the treatment with L-BOAA (A) or 6 hr post transfection with shRNA to Grx1/empty vector (B) and MMP was assessed using JC-1 as the indicator dye.

Loss of MMP caused by L-BOAA (A) and Grx1 downregulation (B) which can be observed as green monomers, is prevented by cyclosporine pretreatment to the cells shown as red aggregates in healthy cells. Bar represents 120 µm.

Figure 10: Basal expression levels of Grx1, Grx2 and TR in SHSY-5Y.
Expression levels of Grx1, Grx2 and TR were checked in SHSY-5Y by immunostaining, cells possess sufficient amount of these oxidoreductases under normal physiological conditions. Bar represents 100 µM.

Figure 11: Estrogen mediated upregulation of Trx1 and Trx2 is prevented by ICI 182,780.

SHSY-5Y cells differentiated with DBA (1 mM) for 24 hr were treated with 17-β estradiol (200 nM) for 24 or 48 hr. Cells were also treated with ICI 182,780 (1 nM) prior to exposing them to estrogen. Upregulation of Trx1 (A) and Trx2 (B) by estrogen is prevented by ICI 182,780.

Figure 12: Estrogen mediated upregulation of TR is prevented by ICI 182,780.
SHSY-5Y cells differentiated with DBA (1 mM) for 24 hr were treated with 17-β estradiol (200 nM) for 24 or 48 hr. Cells were also treated with ICI 182,780 (1 nM) prior to exposing them to estrogen. TR is upregulated by estrogen is prevented by ICI 182,780.

Figure 13: Grx2 is not upregulated by estrogen.
SHSY-5Y cells differentiated with DBA (1 mM; A,B,C,D) for 24 hr were treated with 17-β estradiol (200 nM) for 24 (B) or 48 (C) hr. Cells were also treated with ICI 182,780 (1 nM; D) prior to exposing them to estrogen. Protein levels of Grx2 are not affected by estrogen.

Figure 14: Estrogen upregulates Grx1 and confers protection against L-BOAA toxicity by maintaining MMP.

(A) Immunocytochemical staining of estrogen receptors α and β in SH-SY5Y cells. Bar represents 100 μm.

(B) SH-SY5Y cells were treated with 17-β estradiol (200 nM) for 24 and 48 hr. One set of cells was pretreated with estrogen receptor antagonist ICI 182780 (1 nM), 1 hr prior to the treatment with 17-β estradiol. Immunostaining for Grx1 shows its upregulation on exposure to 17-β estradiol, which is prevented by ICI 182,780. Bar represents 100 μm.

(C) Immunoblot showing upregulation of Grx1 in response to 17-β estradiol (+Est) as compared with control (Con). Densitometric quantitation of immunoblot after normalization with β-tubulin. Data is represented as mean $\pm$ SD from 3 independent experiments. Asterisks indicate values significantly different from controls ($p< 0.05$).

(D) SH-SY5Y cells were exposed to estrogen (200 nM) for 24 hr before treating them with L-BOAA (1 mM; 24 hr) and loaded with JC-1 for monitoring MMP. L-BOAA mediated loss of MMP was abolished in cells pretreated with estrogen as compared to vehicle treated controls. Bar represents 120 μm.

(E) Cells were pretreated with 17-β estradiol (200 nM) or vehicle for 24 hrs before exposure to L-BOAA (500 μM; 24 hr). 17-β estradiol protects against L-BOAA mediated cytotoxicity. Data is represented as mean $\pm$ SD from 3 independent experiments. Asterisks indicate values significantly different from controls ($p< 0.05$).

Figure 15: Exposure to estrogen or overexpression of Grx1 abolishes L-BOAA induced MMP loss in SH-SY5Y cells.

Cells pretreated with 17-β estradiol (200 nM) and cells stably overexpressing Grx1 were treated with L-BOAA (1 mM; 24 hr).

Vehicle treated cells show abrupt decrease in TMRM intensity after CCCP treatment representing sudden loss of MMP (A). Gradual decrease in TMRM intensity in cells treated with L-BOAA represents steady loss of MMP before CCCP treatment, which further decays on adding CCCP (B). MMP is maintained in SH-SY5Y pretreated with 17-β estradiol (C) and in cell lines overexpressing Grx1 (D). L-BOAA mediated loss of MMP is prevented by either pretreating the cells with 17-β estradiol or overexpression of Grx1 (E). Data is represented as mean ±SEM from 3 independent experiments. Asterisks indicate values significantly different from controls ($p < 0.05$). Arrow represents time point of addition of CCCP.

Oxidation of thiol groups of voltage dependent anion channel (VDAC) by Grx1 knockdown: Grx1, a cytosolic enzyme appears critical for maintenance of mitochondrial integrity however the underlying mechanism is unclear. Grx1 is an oxido-reductase and is involved in maintaining the redox status of several redox sensitive proteins. We therefore hypothesized that Grx1 may participate in crosstalk across the mitochondrial membrane through the modification of mitochondrial membrane proteins. VDAC and ANT are components of mPTP that are present in the outer and inner mitochondrial membranes, respectively. They are redox sensitive proteins and can be potentially modulated by Grx1 by altering the redox status of their thiol groups. Neuro-2a cells were transfected with empty vector or shRNA to Grx1 and cells were collected after 72 hr and incubated (derivatized) with AIS, an alkylating agent that binds to free thiol groups in proteins. AIS derivatized samples were separated by electrophoresis under non-reducing conditions and subjected to immunoblotting using antibody to VDAC and ANT. We observed significantly lower amounts of AIS derivatized (reduced) VDAC in cells wherein Grx1 was knocked down as compared to cells transfected with vector alone (Fig. 16A), while total VDAC levels (reduced plus oxidized) were unchanged indicating that VDAC protein was oxidatively modified by Grx1 knockdown. Moreover, ANT, an inner membrane mitochondrial protein did not show oxidative modification upon downregulation of Grx1 and the reduced state of ANT was increased significantly (Fig. 16B), indicating that Grx1 downregulation oxidatively modifies the redox state of outer membrane but not inner membrane proteins in the mitochondria (see discussion on page 252). Cells were also transfected with construct harbouring scrambled to Grx1 and levels of reduced and total VDAC compared among empty vector/scrambled transfected samples. No difference in total and reduced VDAC levels was observed while using either mU6pro empty vector or scrambled sequence as control for transfections (Fig. 17).

Figure 16: Downregulation of Grx1 causes oxidative modification of thiol groups of voltage dependent anion channel (VDAC).

Neuro-2a cells were transfected with shRNA to Grx1 or empty vector and the cell lysates were incubated with AIS which alkylates the free thiol groups thus causing a shift in the migration on non-reducing SDS-PAGE.

(A) Immunoblot depicting reduced and AIS derivatized VDAC. Total cell lysate of Neuro-2a cells transfected with empty vector (shRNA '-') or shRNA to Grx1 (shRNA '+') were subjected to non-reducing SDS-PAGE. Reduced VDAC measured as such (AIS '-') and as AIS derivatized VDAC (AIS '+'), is shown. Total VDAC from control (shRNA '-') or shRNA to Grx1 (shRNA '+') transfected cells was measured using reducing SDS-PAGE followed by immunoblotting.

(B) Densitometric measurements of the immunoblots depicted in (A). Reduced and AIS derivatized VDAC were normalized with total VDAC. Values are mean $\pm$ SD (n=6) individual experiments. Asterisks indicate values significantly different from controls (p< 0.01).

(C) Immunoblot depicting reduced and AIS derivatized ANT.
Total cell lysate of Neuro-2a cells transfected with empty vector (shRNA '-') or shRNA to Grx1 (shRNA '+') were subjected to non-reducing SDS-PAGE. Reduced ANT measured as such (AIS '-') and as AIS derivatized ANT (AIS '+'), are depicted. Total ANT from control (shRNA '-') or shRNA to Grx1 (shRNA '+') transfected cells was measured using reducing SDS-PAGE followed by immunoblotting.

(E) Densitometric measurements of the immunoblots shown in (C). Reduced and AIS derivatized ANT were normalized with total ANT. Values are mean $\pm$ SD (n=6) individual experiments. Asterisks indicate values significantly different from controls (p< 0.01).

Figure 17: Scrambled contruct to Grx1 shRNA or empty vectror did not alter either levels of Grx1 or redox status of VDAC.

In cells transfected with empty vector or scrambled construct, the expression of Grx1 as assessed by qRT-PCR showed no change, however in cells transfected with shRNA to Grx1, significant decrease (56%) was observed. Data is mean ± SD (n=5 experiments).

B- Neuro-2a cells were transfected with empty vector or scrambled shRNA and the cell lysates were incubated with AIS which alkylates the free thiol groups thus causing a shift in the migration on non-reducing SDS-PAGE. The redox status of VDAC remained unaltered when cells were transfected with empty vector or scrambled shRNA as shown by the reduced VDAC measured as such and as AIS derivatized VDAC, although it reduced significantly when cells were transfected with shRNA to Grx1 (data shown earlier).

2) Loss of DJ-1 by knockdown of glutaredoxin (Grx1), triggers translocation of Daxx and ensuing cell death.

DJ-1 is a redox sensitive protein and its loss of function mutations are implicated in pathogenesis of PD. To investigate if its redox modulation by increased oxidative condition impairs its function which may be happening in sporadic cases of PD, effect on DJ-1 and downstream events were studied in response to Grx1 knockdown in Neuro-2a cells. To begin with this study, the Grx1 knockdown model was characterized again to verify the knockdown and generation of reactive oxygen species (ROS).

Knockdown of glutaredoxin 1 generates reactive oxygen species: Grx1 was downregulated in neuroblastoma cells of mouse origin, (Neuro-2a) by transiently transfecting them with shRNA to Grx1 and the knockdown was verified by immunocytochemistry (Fig. 18A; GFP positive cells represent transfected cells; Grx1 labeled by Alexa 594), quantitative real time PCR (Fig. 18B), and by immunoblotting (Fig. 18C). Whereas Grx1 was consistently silenced by 50-60 % using shRNA to Grx1, levels of Grx1 were unaffected when cells were transfected with either empty vector (control) or by using scrambled sequence (Fig. 18B). Grx1 knockdown resulted in increased generation of reactive oxygen species as observed by H_2DCFDA (2',7'-dichlorodihydrofluorescein diacetate) staining (Fig. 18D) and quantitative fluorimetric assay of intracellular ROS (Fig. 18E).

Downregulation of Grx1 leads to decrease in protein levels of DJ-1 whereas mRNA levels were unaltered: Grx1 helps in regulating the intracellular redox milleu and its downregulation generates ROS. Since DJ-1 is a redox sensitive protein and mutations in it have been implicated in the pathogenesis of Parkinson's disease we studied the effect of

Figure 18: Knockdown of Grx1 using shRNA leads to generation of reactive oxygen species.

Neuro-2a cells transfected with pAAV-GFP (Control) or pAAV-GFP-shRNA (shRNA-Grx1) were processed for immunostaining for Grx1, 72 hr post transfection. (A) GFP expression represents transfected cells which show decreased expression of Grx1 in shRNA-Grx1 (lower panel) as compared to the GFP expressing control (upper panel). Bar represents 100 µm. (B) Neuro-2a cells transfected with either mU6-shRNA-Grx1 or scrambled or empty vector were collected 72 hr later and processed for quantitative real time PCR and immunoblotting. shRNA to Grx1 resulted in 50% decrease in mRNA levels of Grx1 as compared to the empty vector and scrambled sequence which did not show any significant change (n=4). (C) Representative immunoblot showing decrease in protein levels of Grx1. β-tubulin levels were measured as loading control. Densitometric analysis of the immunoblots representing the relative intensity of the immunoreactive bands, showing ~60% decrease of Grx1 in cells transfected with shRNA-Grx1 as compared to the empty vector transfected controls (n=6). (D-E) Neuro-2a cells transfected with mU6-shRNA-Grx1 or empty vector were either loaded with 10 µM H_2DCFDA for 15 min and imaged to examine DCF fluorescence or incubated with 10 µM H_2DCFDA for 45 min and processed to quantitate DCF fluorescence using fluorimetry (n=9). shRNA to Grx1 enhances ROS generation as observed by (D) imaging or (E) quantitated by fluorimetry. Values are mean ± SD. Asterisks indicate values significantly different from empty vector transfected controls (P<0.001).

Grx1 downregulation on DJ-1. Knockdown of Grx1 by shRNA resulted in decrease in protein levels of DJ-1 in Neuro-2a cells, as observed by immunocytochemistry (Figure 19A), and by immunoblotting (Fig. 19B) as opposed to when empty vector or scrambled sequence was used. In order to ascertain if the downregulation of DJ-1 was at transcriptional level, we examined its mRNA expression levels. Grx1 silencing did not result in transcriptional downregulation of DJ-1 as inferred from quantitative real time PCR (Fig. 19C).

Depletion of glutathione (GSH) using BSO also elicits oxidative stress by generation of ROS and loss of mitochondrial membrane potential (MMP): To study whether the effects of Grx1 knockdown downstream to DJ-1 depletion were specific to Grx1 downregulation or due to global oxidative stress, a parallel paradigm to generate oxidative stress was used. Depletion of GSH using L-buthionine-S,R-sulfoximine (BSO), a γ-glutamyl cysteine synthetase inhibitor, is known to generate reactive oxygen species (Merad-Boudia et al., 1998; Gabryel and Malecki, 2006). GSH depletion using BSO augmented oxidative stress by inducing ROS generation and loss of MMP in Neuro-2A cells. Neuro-2a cells were treated with different concentrations of BSO (50, 100, 200, 300 and 500 µM) or vehicle (0.9% saline) for 48 hr and loss of MMP was studied. Treatment of Neuro-2a with BSO resulted in loss of MMP (Fig. 20A) in a dose response manner as observed by JC-1 staining. The loss was optimum (approximately 60%) when 100 µM of BSO was used, hence this dose for used for further experiments. Further, it also augmented the generation of ROS (Fig. 20B) thereby causing oxidative stress, an effect similar to what was observed subsequent to Grx1 downregulation. Hence, GSH depletion using BSO was used as a parallel paradigm to study events downstream to DJ-1 depletion resulting due to Grx1 downregulation.

Figure 19: Grx1 knockdown reduces DJ-1 protein levels; mRNA remains unaltered.

Neuro-2a cells transfected with pAAV-GFP (Control) or pAAV-GFP-shRNA (shRNA-Grx1) were processed for immunostaining for DJ-1, 72 hr post transfection. (A) GFP expression represents transfected cells, which show decreased expression of DJ-1 in cells transfected with shRNA-Grx1 (lower panel) as compared to the GFP expressing control (upper panel). Bar represents 100 µm. (B-C) Neuro-2a cells transfected with either mU6-shRNA-Grx1 or scrambled/empty vector were collected 72 hr later and processed for immunoblotting and quantitative real time PCR. (B) Representative immunoblot showing decrease in protein levels of DJ-1 in cells transfected with shRNA to Grx1; scrambled showed no change w.r.t. empty vector transfected control. β-tubulin levels were measured as loading control. Densitometric analysis of the immunoblots representing the relative intensity of the immunoreactive bands, showing ~25% decrease of DJ-1 in cells transfected with shRNA-Grx1 (n=5). (C) shRNA to Grx1 and scrambled sequence did not cause any change in mRNA levels of DJ-1 as compared to the empty vector transfected cells. Values are mean ± SD. Asterisk wherever indicated designates values significantly different from empty vector transfected controls (P<0.001).

Figure 20: GSH depletion augments ROS generation and loss of MMP, an effect elicited by Grx1 knockdown also.

Neuro-2a cells treated with 100 µM BSO or vehicle for 48 hr were processed for ROS measurement and qualitative determination of loss of MMP. For quantitative measurement of ROS, cells were incubated with 10 µM H$_2$DCFDA for 45 min at the termination of BSO treatment, collected and processed for fluorimetric assay. (A) Cells treated with BSO show increased levels of ROS as compared to the vehicle (0.9% saline) treated controls. Values are mean ± SD. Asterisk depicts values significantly different from controls (P<0.001). (B) MMP is lost with increasing dose of BSO as seen by accumulation of diffused JC-1 stain in cells.

Silencing of Grx1 elicit oxidative stress and induce upregulation of stress responsive genes: Stress elicited by an oxidative insult is regulated by intracellular compensatory mechanisms as reported earlier (Papadia et al., 2008). To investigate if downregulation of Grx1 induces the expression of stress responsive genes we examined the alterations in expression level of proteins responsible for maintaining redox homeostasis. Thioredoxin system is well acknowledged for maintaining redox status and cytosolic superoxide dismutase 1 (SOD1) defends the cells against oxidative damage by free oxygen radical. With this rationale, lysates from Neuro-2a cells, transfected with either empty vector (control) or with shRNA to Grx1 were subjected to SDS-PAGE and levels of thioredoxin 1 (Trx1) and superoxide dismutase (SOD1) were examined. Protein levels of SOD1 (Fig. 21A) and Trx1 (Fig. 21B) were elevated in shRNA transfected samples as compared to the control.

Translocation of DJ-1 to mitochondria is specific to Grx1 downregulation and does not occur after GSH depletion: DJ-1 is normally localized in cytosol and nucleus but translocates to mitochondria following oxidative insult where it implements its neuroprotective actions. To explore the subcellular localization of DJ-1 in response to decreased levels of Grx1, Neuro-2a cells were transfected with shRNA to Grx1 or empty vector, loaded with MitoTracker and immunostained for DJ-1. DJ-1 protein levels were decreased in cells transfected with shRNA to Grx1 and it colocalized with MitoTracker suggesting its localization in mitochondria (~90% cells -mitochondrial; 17% cells - nuclear) whereas DJ-1 was localized in cytosol and nucleus in cells transfected with empty vector [~80% cells - nuclear; 10% cells - mitochondrial (Fig. 22A)]. To explore if translocation of DJ-1 to mitochondria is an outcome of general oxidative stress or specific

Figure 21: Grx1 knockdown trigger upregulation of stress responsive proteins.

Neuro-2a cells transfected with either shRNA to Grx1 or empty vector were collected 72 hr later for assessing protein levels of SOD1 and Trx1. (A-B) Downregulation of Grx1 using shRNA resulted in upregulation of both (A) SOD1 and (B) Trx1. Representative immunoblot showing increased protein levels of SOD1 and Trx1 in cells transfected with shRNA to Grx1 as compared to empty vector transfected control. Densitometric analysis of the immunoblots representing the relative intensity of the immunoreactive bands, showing increase in the levels of both SOD1 and Trx1 in cells transfected with shRNA-Grx1 (n=9 & 12 respectively). Values are mean ± SD. Asterisk wherever indicated, designate values significantly different from controls (P<0.001).

to Grx1 knockdown, Neuro-2a cells were treated with either BSO or vehicle and processed similarly to study the localization of DJ-1. Approximately 70% of the cells either treated with saline or BSO were found to have DJ-1 localized in the nucleus whereas 20% cells had DJ-1 present in the mitochondria irrespective of the treatment (Fig. 22B) suggesting that GSH depletion does not trigger translocation of DJ-1 to mitochondria. This implies that translocation of DJ-1 was specifically elicited by downregulation of Grx1 and not by oxidative stress generated by depletion of glutathione.

Silencing of Grx1 results in translocation of Daxx to the cytosol and induces cytotoxicity, a phenomenon not triggered by GSH depletion: DJ-1 confines Daxx, the death repressor protein, within the nucleus during normal physiological conditions, which gets released into the cytosol in response to an oxidative stimulus where it propagates the cell death cascade. Downregulation of Grx1 leads to reduction in the protein levels of DJ-1 in the nucleus. To infer whether the decreased levels of DJ-1 affects Daxx and causes subsequent toxicity, we studied its localization within the nucleus. Neuro-2a cells transfected with shRNA to Grx1 showed translocation of Daxx (~ 87 % cells with respect to control) from the nucleus to the cytosol (Fig. 23B) as compared to the cells transfected with empty vector (control), wherein it was retained within the nucleus. Grx1 downregulation also resulted in cell death (~60%) as compared to the empty vector transfected control (~20%), as inferred by TUNEL assay (Fig. 23A). To understand if Daxx translocation and cytotoxicity is a consequence of general oxidative stress stimuli or specific to depletion of Grx1 we explored the localization of Daxx and subsequent cytotoxicity following treatment with BSO.

Figure 22: Grx1 downregulation and not GSH depletion causes mitochondrial translocation of DJ-1.

Neuro-2a cells either transfected with shRNA to Grx1/empty vector or treated with BSO/vehicle were loaded with MitoTracker 3 hr prior to fixation, immunostained for DJ-1 post fixation and imaged using confocal microscope (LSM50). (A) Grx1 knockdown using shRNA resulted in decreased levels of cytosolic and nuclear DJ-1 and its translocation to mitochondria unlike to the cells transfected with empty vector where it was confined within the nucleus and cytosol. Bar represents 10 µm. Quantification represents percentage of cells showing nuclear retention and mitochondrial translocation of DJ-1 per total number of cells as represented by MitoTracker staining (n=4). Asterisk indicates values significantly different from empty vector transfected controls (P=0.003). (B) BSO (100 µM) or vehicle treatment did not trigger mitochondrial translocation of DJ-1 as illustrated by the representative image and quantitation (n=4). Bar represents 10 µm. Values are mean ± SD.

Contrastingly to what was observed following Grx1 downregulation, depletion of GSH using BSO in Neuro-2a cells, did not trigger either cytosolic translocation of Daxx (Fig. 23D) or cytotoxicity (Fig. 23C) as compared to vehicle (saline) treated controls, thus suggesting that translocation of Daxx and consequential cytotoxicity was specifically caused by silencing of Grx1 and was not an outcome of general oxidative stress stimuli.

Exogenously expressed wild type (WT) DJ-1 protein levels, although reduced following Grx1 downregulation, could still prevent cytosolic translocation of Daxx and attenuate cytotoxicity: To explore if overexpressed WT DJ-1 could potentially prevent the cytotoxicity caused by downregulation of Grx1, Neuro-2a cells were co-transfected either with a combination of shRNA to Grx1 and pcDNA, or with wild type DJ-1 overexpressing construct along with either shRNA to Grx1 or scrambled sequence. Grx1 downregulation resulted in decline in protein levels of endogenous as well as exogenously expressed DJ-1, strikingly, the decrease observed in exogenously expressed DJ-1 was more prominent as compared to that of endogenous DJ-1 (Fig. 24A), however, overexpressing wild type DJ-1 attenuated the cytotoxicity caused by Grx1 downregulation (Fig. 24B). To investigate if cytotoxicity observed by Grx1 downregulation was mediated by translocation of Daxx and downstream mechanisms, we studied its localization in cells transfected with shRNA to Grx1 and wild type DJ-1. The cytosolic translocation of Daxx caused by Grx1 downregulation was completely eliminated by overexpressing wild type DJ-1 (Fig. 24C) suggesting that Grx1 knockdown induced cytosolic translocation of Daxx and triggered other mechanisms as well, which lead to cytotoxicity.

Figure 23: Daxx translocation and cytotoxicity are triggered by Grx1 knockdown but not by GSH depletion.

Neuro-2a cells either transfected with shRNA to Grx1/empty vector or treated with BSO/vehicle were fixed 72 and 48 hr later respectively, and processed either for TUNEL assay or immunostaining for Daxx. (a-b) Cells transfected with shRNA to Grx1 caused significant (a) cytotoxicity and (b) cytosolic translocation of Daxx as compared to respective empty vector transfected controls. Quantification represents (a) percentage of cells positive for TUNEL staining per total number of cells as represented by DAPI staining and (b) percentage of cells showing cytosolic translocation of Daxx per total number of cells as represented by PI staining (n=4). Values are mean ± SD. Asterisk indicate values significantly different from empty vector transfected controls (P<0.001). (c-d) BSO treatment to cells (c) neither induced cytotoxicity (d) nor triggered cytosolic translocation of Daxx similar to vehicle treated controls (n=6). Values are mean ± SD. Bar represents 200 μm for A & C, 5 μm for B and 10 μm for D.

Figure 24: Exogenously expressed DJ-1 attenuates cytotoxicity and prevents Daxx translocation inspite of its protein levels being diminished by Grx1 knockdown.

Neuro-2a cells were co-transfected with DJ-1 along with shRNA to Grx1 or scrambled construct. (A-C) For silencing Grx1, cells were co-transfected with shRNA to Grx1 and pcDNA, whereas control cells were co-transfected with scrambled and pcDNA (all in the ratio of 1:1) and processed for immunoblotting, TUNEL assay and immunostaining. (A) Grx1 downregulation using shRNA depletes protein levels of exogenously expressed DJ-1. Representative immunoblot showing reduced protein levels of exogenously expressed DJ-1 in cells co-transfected with shRNA to Grx1. β-tubulin levels were measured as loading control. Densitometric analysis of the immunoblots representing the relative intensity of the immunoreactive bands, showing more prominent decrease in the levels of exogenous DJ-1 (n=6). (B-C) Overexpression of DJ-1 (B) attenuates cytotoxicity and (C) prevents translocation of Daxx mediated by shRNA to Grx1. Quantification represents (B) percentage of cells positive for TUNEL staining per total number of cells as represented by DAPI staining and (C) percentage of cells showing cytosolic translocation of Daxx per total number of cells as represented by PI staining (n=6 & 8 respectively). Values are mean ± SD. Asterisk wherever indicated designates values significantly different from scrambled transfected controls (P<0.001).

Protein levels of exogenously expressed DJ-1 mutants were unaltered by Grx1 silencing and did not protect against cytotoxicity mediated by Grx1 knockdown: To examine if redox modification of DJ-1 following Grx1 knockdown is responsible for decrease in its protein levels, we co-transfected Neuro-2a cells with different mutants of DJ-1 such as C106A, C53A, L166P along with either shRNA to Grx1 or scrambled construct. Grx1 downregulation did not cause decrease in the protein levels of either of the mutants of DJ-1, unlike that observed with wild type DJ-1 (Fig. 25A), thus suggesting indirectly, that Grx1 was helping in maintenance of the redox status of cysteine residues present in it and DJ-1 mutants wherein cysteine has been replaced by alanine were not affected by Grx1 downregulation. However, protein levels of recombinant L166P were very low (Fig. 25A) similar to that reported earlier (Moore et al., 2003). L166P, due to its highly unstable conformation, rapidly undergoes proteasome degradation (Miller et al., 2003; Olzmann et al., 2004). To investigate if the mutants of DJ-1 potentially offered protection against cytotoxicity mediated by Grx1 silencing, Neuro-2a cells co-transfected with WT or with either of the mutants of DJ-1 along with shRNA to Grx1 or scrambled sequence were subjected to TUNEL assay. Only WT DJ-1 could attenuate the cytotoxicity remarkably unlike mutants of DJ-1 (Fig. 25B). L166P caused massive cytotoxicity by itself and along with shRNA to Grx1, the toxicity was further exacerbated leading to loss of cells during the experiment thus rendering it technically impossible to count TUNEL positive cells.

DJ-1 is not modulated by Grx1 through direct interaction: We could not detect a direct interaction of Grx1 with DJ-1 by co-immunoprecipitation (Fig. 26A & B) thus indicating that Grx1 modulates the redox status of DJ-1 indirectly and maintains its stability. Co-immunoprecipitation was done using antibody against FLAG for immunoprecipitation and the blot was probed for Grx1 (Fig. 26A). A reverse co-immunoprecipitation was also done using antibody against Grx1 for immunoprecipitation and the blot was probed for DJ-1 (Fig. 26B). No interaction was observed in either case suggesting that Grx1 was modulating the redox state of DJ-1 indirectly. A negative control was also run wherein anti mouse IgG was used instead for immunoprecipitation (Fig. 26 A & B).

Thiol groups of DJ-1 are oxidatively modified in response to Grx1 knockdown: Grx1 potentially modulates redox status of protein thiol groups and its knockdown can oxidatively modify the thiol groups (Saeed et al., 2008). To verify if oxidative modification of thiol groups is the causative factor of loss of DJ-1 protein, we looked at the redox status of exogenously expressed wild type FLAG DJ-1. Neuro-2a cells were co-transfected with wild type FLAG DJ-1 along with either scrambled or shRNA to Grx1, collected 72 hr later and incubated (derivatized) with AMS (30 mM), an alkylating agent that binds to free thiol groups. AMS derivatized samples were separated by electrophoresis under non-reducing conditions and subjected to immunoblotting using antibody to FLAG. Both total FLAG DJ-1 and AMS derivatized FLAG DJ-1 signals were significantly reduced in cells wherein Grx1 was knocked down when compared to controls (Fig. 27A), and also the ratio of AMS derivatized to total FLAG DJ-1 did not change in cells showing Grx1 knockdown or scrambled transfected control groups (Fig. 27B), indicating that the oxidatively modified DJ-1 was lost soon after modification and the intact protein left was essentially present in the reduced form.

Figure 25: Exogenously expressed DJ-1 mutants are not affected by Grx1 knockdown and fail to prevent ensuing cytotoxicity.

Neuro-2a cells were co-transfected with WT or mutants of DJ-1 (C106A, C53A, L166P) along with shRNA to Grx1 or scrambled construct, collected 72 hr later and processed for immunoblotting or TUNEL assay. (A) Downregulation of Grx1 did not affect the protein levels of mutants of DJ-1 but WT DJ-1 levels were decreased. Representative immunoblot probed with antibody to DJ-1 (left blot) and anti-FLAG antibody (right blot) showing similar protein levels of mutants of DJ-1 in cells transfected with either shRNA or scrambled construct but reduced levels of exogenous WT DJ-1 in cells transfected with shRNA to Grx1. β-tubulin levels were measured as loading control. Densitometric analysis of respective immunoblots representing the relative intensity of the immunoreactive bands, showing decrease in the levels of exogenous WT DJ-1 but no effect on mutants. (B) Mutants of DJ-1 do not protect from cytotoxicity caused by downregulation of Grx1. Quantification represents percentage of cells positive for TUNEL staining per total number of cells as represented by DAPI staining. Values are mean $\pm$ SD (n=4 for A & B). Asterisk represents significantly different values wherever indicated (P<0.001).

Figure 26: Grx1 does not interact with DJ-1 directly.

Neuro-2a cells were transfected with constructs overexpressing human Grx1 and FLAG tagged DJ-1 and the lysates were immunoprecipitated with antibody against (A) FLAG and (B) Grx1. Lanes indicate the following:- M-Marker, Input, IP-F-immunoprecipaitation using antibody to FLAG, Sup-20 µl of supernatant left after IP with FLAG antibody, IP-I-immunoprecipaitation using anti mouse IgG, Supn-20 µl of supernatant left after IP with anti mouse IgG, Grx++-positive control (Overexpressed Grx1) for blot (A); For blot (B) IP-G indicates mmunoprecipaitation using antibody to Grx1.

Figure 27: Grx1 downregulation causes oxidative modification of wild type FLAG DJ-1.

Neuro-2a cells were co-transfected with wild type FLAG DJ-1 along with either shRNA to Grx1 or scrambled and the cell lysates were incubated with AMS which alkylates the free thiol groups thus causing a shift in the migration on non-reducing SDS-PAGE. (A) Immunoblot depicting total FLAG DJ-1 and tubulin in cells transfected with either scrambled or shRNA subjected to reducing SDS-PAGE (upper blot), whereas reduced and AMS derivatized FLAG subjected to non-reducing SDS-PAGE (lower blot). Densitometric measurements of the immunoblots depicted on the left. Total and AMS derivatized signals were normalized with β-tubulin levels. (B) Total FLAG signals were normalized with β-tubulin and AMS derivatized FLAG signals were normalized with total FLAG. Values are mean ± SEM (n=6) individual experiments. Asterisks indicate values significantly different from controls (P<0.01).

Proteasome degradation is not responsible for depletion of endogenous DJ-1 following Grx1 knockdown: DJ-1 does not undergo degradation via ubiquitin proteasome pathway normally as has been shown earlier (Gorner et al., 2004). Our experiments also revealed that endogenous DJ-1 does not accumulate by inhibiting proteasome using MG132 although β-catenin starts accumulating within an hour and shows a peak accumulation at 6 hr (Fig. 28A). To investigate if oxidative modification of DJ-1 induced by Grx1 knockdown makes it susceptible to proteasome degradation, Neuro-2a cells were transfected with either scrambled construct or shRNA to Grx1 and treated with proteasome inhibitor, MG132 (10 µM) for 0 or 7 hr prior to harvesting. Levels of DJ-1 decreased 48 hr following transfection and no accumulation of DJ-1 was found in cells treated with MG132 for 0 or 7 hr in either scrambled or shRNA transfected cells thus demonstrating that proteasome mediated degradation is not responsible for depletion of DJ-1 following Grx1 knockdown (Fig. 28B). To examine the mode of degradation of exogenously expressed DJ-1, Neuro-2a cells were co-transfected with FLAG tagged WT DJ-1 along with shRNA to Grx1 or scrambled construct followed by inhibition of proteasome machinery using MG132 (10 µM) for 7 hr following 41 hr post-transfection. Strikingly, unlike endogenous DJ-1, depletion of ectopically expressed FLAG tagged WT DJ-1 was partially prevented by treating cells with MG132 as opposed to vehicle (DMSO). To confirm the levels of DJ-1 prior to treatment, untreated cells were collected 41 hr post transfection (NT; Fig. 28C). Accumulation of β-catenin levels in cells treated with MG132 for 7 hr confirmed inhibition of proteasome as it undergoes proteasome mediated degradation as reported earlier (Kim et al., 2009).

Figure 28: Depletion of exogenously expressed FLAG tagged WT DJ-1 and not endogenously expressed one is partially caused by proteasome degradation machienery.

Neuro-2a cells (A) were treated with MG132 (10 μM) after growing them for 36 hr and levels of endogenous DJ-1, tubulin and β-catenin were examined. Cells were also transfected with (B) scrambled/shRNA-Grx1, and treated with MG132 (10 μM) for 0 or 7 hr prior to harvesting, i.e. 48 hr following transfection. Grx1 knockdown results in loss of DJ-1 protein levels in cells transfected with shRNA. Accumulation of DJ-1 is not seen following treatment with MG132 in either scrambled or shRNA transfected cells whereas β-catenin accumulates following 7 hr of treatment with MG132 (n=4; P=<0.001). (C) Neuro-2a cells co-transfected with FLAG tagged WT DJ-1 and shRNA or scrambled construct were treated with MG132 or DMSO (vehicle) 41 hr post transfection for 7 hr. Non treated (NT) cells were collected at 41 hr proceeding transfection. Loss of exogenous DJ-1 caused by shRNA to Grx1 is partially prevented by MG132 along with β-catenin accumulation after 7 hr treatment (n=3; p<0.05). Values are mean ± SD. Asterisk indicate values significantly different from scrambled transfected controls.

3) Mitochondrial Glutaredoxin (Grx2) confers protection against MPP$^+$ mediated cell toxicity.

Grx2 is a mitochondrial glutaredoxin, like Grx it is also a thiol/disulfide oxidoreductase component of GSH system and is involved in the reduction of GSH based mixed disulfide in the mitochondria (Gladyshev et al., 2001). It helps in maintaining redox homeostasis and participates in a variety of cellular redox pathways. Since it is an important redox regulator within the mitochondria, there is a need to exploit its potential as a redox regulating enzyme.

Subcellular localization of Grx2 in mitochondria: To understand its role in affording protection against MPP$^+$ mediated cytotoxicity, Grx2 was overexpressed in neuroblastoma cells. Specificity of the antiserum used for confirming the overexpression was examined. Neuro-2a cells loaded with MitoTracker prior to fixation were immunostained with antiserum to Grx2, post fixation. Grx2 co-localized with MitoTracker indicating its localization in mitochondria thereby confirming the specificity of the antiserum (Fig. 29).

Overexpression of Grx2 abolishes MPP$^+$ mediated cytotoxicity in Neuro-2a cells: Neuro-2a cells were transfected with pCMV-Grx2 (Figs. 30 and 31B) or empty vector (control) and then exposed to MPP$^+$ (1 mM) for 24 h, as this dose was found appropriate after performing a dose response (Fig. 31A). Cell death was assessed using the terminal deoxynucleotidyl transferase biotin-dUTP nick end labeling (TUNEL). Cells transfected with empty vector when exposed to MPP$^+$ showed significantly higher TUNEL immunoreactivity (37±5.7% of cells) as compared to cells exposed to vehicle (7.7±2.5%). Grx2 overexpressing cells exposed to MPP$^+$ revealed less TUNEL immunoreactivity (11.7± 3.2%; Fig. 31B) indicating the lowered levels of cell death as compared to control cells exposed to MPP$^+$. Grx2 overexpressing cells (6.8±4.5%) had similar number of apoptotic nuclei as cells transfected with empty vector.

Figure 29: Localization of Grx2 in mitochondria.

Neuro-2a cells loaded with MitoTracker were immunostained with antiserum to Grx2 and visualized to examine the sub-cellular localization of Grx2. Immunocytochemical detection of Grx2 is depicted in green, while staining with MitoTracker is shown in red in Neuro-2a cells. MitoTracker staining colocalizes with Grx2 indicating the specificity of the antiserum used (scale bar depicts 10 μm). The lower panel shows a confocal image of a cell with the mitochondria highlighted using MitoTracker which colocalizes with Grx2 (Scale bar depicts 5 μm).

Figure 30: Overexpression of Grx2 in Neuro-2a cells.

Neuro2a cells were transfected with pCMV-Grx2/empty vector and overexpression verified by immunoblotting and immunostaining. (A) Immunostaining for Grx2 in N2A cells transfected with empty vector (pCMV-MCS; Control) or with pCMV-MCS containing the complete ORF of Grx2 (Grx2) is shown along with respective DAPI images (Scale bar represents 100 μM); Representative immunoblot stained with Grx2 antiserum of control and Grx2 overexpressing cells is depicted above.

Figure 31: Overexpression of Grx2 abolishes MPP+-mediated toxicity in Neuro2a cells. (A) Neuro-2a cells were treated with varying concentration of MPP$^+$ to standardise the optimum dose as shown by the dose response curve and 1 mM dose was found optimum. (B) Cells were exposed to MPP$^+$ (1 mM) for 24 h and the percentage of TUNEL-positive cells vs. DAPI-stained nuclei was estimated from 3 independent experiments. Asterisk indicates values significantly different from controls (p<0.001). Images of TUNEL positive cells along with corresponding DAPI images are depicted (scale bar represents 200 µM).

II. **Mechanisms of cell death in Parkinson's disease and role of estrogen.**

1) MPP$^+$ induced activation of p38 in primary culture of human dopaminergic neurons.

Primary neurons derived from human progenitor cells show dopaminergic phenotype: Primary cultures of human CNS progenitor cells were used to study altered MAP Kinase mediated apoptotic signalling in response to MPP$^+$. The cells were initially characterized for their lineage and expression of different molecules involved in apoptotic signalling. Primary cultures of human CNS progenitor cells were differentiated into neurons and characterized for the dopaminergic phenotype by studying the expression of tyrosine hydroxylase (TH) and neuronal marker β-III-tubulin (Tuj1) (Fig. 32A & B; low and high magnification). More than 85% cells were found to be positive for both TH and Tuj1 and could be co-localized (Fig. 31C). On average, 89.9% cells were found to be positive for Tuj1 and 87.7% of the DAPI stained nuclei were found to be positive for tyrosine hydroxylase staining (Fig. 32D).

Expression of redox regulating proteins and molecules involved in MAPK mediated apoptotic and survival pathway in primary culture of dopaminergic neurons derived from human progenitor cells: Less than 2% cells stained positive for GFAP thus indicating the homogeneity of neuronal population. Positive immunostaining was also seen for thioredoxin 1, glutaredoxin 1 and redox sensor DJ-1. Immunostaining for estrogen receptor α, β and dopamine transporter was also observed. Proteins involved in survival and cell death pathway such as Akt, PTEN, Fox2, p53, JNK, p38, Ask, Daxx and Bax were also found to be expressed in these cells (Fig. 33).

Figure 32: Characterization of primary dopaminergic neurons derived from human progenitor cells.
Primary cultures of human CNS progenitor cells were isolated, cultured and differentiated into neurons by treating them with appropriate growth factors for three weeks. Cells were then seeded as described earlier and fixed before staining. (A) Immunocytochemistry showing the co-expression of tyrosine hydroxylase (TH), β-III-tubulin (Tuj1) in primary neurons derived from human CNS progenitor cells and the total cell number depicted by DAPI. Bar represents 200 µm. (B) same at higher magnification, bar represents 100 µm. (C) Cells showing co-localization of TH & Tuj1, bar represents 200 µm. (D) Quantification of percentage of cells positive for TH and Tuj1 immunostaining per total number of cells is represented by DAPI count. Data is represented as mean + SD of 4 independent experiments. Asterisks indicate values significantly different from controls (p < 0.05).

Expression of redox regulating proteins and molecules involved in MAPK mediated apoptotic and survival pathway in primary culture of dopaminergic neurons derived from human progenitor cells: Positive immunostaining was also seen for thioredoxin 1, glutaredoxin 1 and redox sensor DJ-1. Immunostaining for estrogen receptor α, β and dopamine transporter was also observed. Proteins involved in survival and cell death pathway such as Akt, PTEN, Fox2, p53, JNK, p38, Ask, Daxx and Bax were also found to be expressed in these cells (Fig. 33).

Dose response for MPP$^+$ and MPTP in primary culture derived from human CNS progenitor cells: MPP$^+$ was used as a model neurotoxin to study altered signaling in these cultures. Cells were responsive to MPP$^+$ mediated toxicity as examined by TUNEL assay and dose-dependent increase in TUNEL positive neurons was observed after exposure to increasing concentration of MPP. A dose response was done to standardize the dose for MPP$^+$ and MPTP both (Fig. 34) and 10 μM MPP$^+$ was found to be the appropriate dose for intersecting pathway of cell death in response to this toxin. MPP$^+$ showed a linear increment in percentage of TUNEL posistive cells with increasing dose concentration [approximately 50% TUNEL positive cells with 10 μM MPP$^+$ (Fig. 34A)] whereas cell death observed with MPTP was less and aberrant with increasing dose concentration (Fig. 34B). Since MPP$^+$ showed 50% cell death at 10 μM concentration, the dose was found appropriate for further experiments.

Figure 33: Expression of redox regulating, apoptotic and survival proteins in primary dopaminergic culture derived from human CNS progenitor cells.

Proteins depicted as labels were found to be expressed in primary dopaminergic culture derived from human CNS progenitor cells. These include dopaminergic neuronal markers such as Tuj1, TH; glial marker-GFAP; oxidoreductases such as Grx1, Trx1; Estrogen receptor α, β, dopamine transporter; cell survival & apoptotic proteins such as Akt, PTEN, Fox2, p53, JNK, p38, Ask, Daxx and Bax. Bar represents 100 μM.

Figure 34: Dose response for MPP⁺ and MPTP toxicity in primary neurons.
Cells were treated with vehicle, MPP⁺ or MPTP at 1, 10 and 100 μM concentration for 24 hr and cell toxicity was verified by TUNEL assay. Bar represents 200 μm. Quantification of percentage of TUNEL positive cells per total number of cells represented by DAPI count at different concentration of MPP⁺ and MPTP. Data is represented as mean ± SD of 4 independent experiments. Asterisks indicate values significantly different from controls ($p < 0.05$).

p38 inhibitor (SB239063) afford protection against MPP⁺ induced cell death in primary neurons derived from human CNS progenitor cells: To investigate if MPP⁺ toxicity is mediated via MAPKKK pathway by activating p38 MAPK, primary dopaminergic cultures derived from human CNS progenitor cells were treated with p38 inhibitor (SB239063) prior to treatment with 10 μM MPP⁺. p38 inhibitor (1 μM), which inhibits both p38 α and β, conferred protection to these cells against MPP⁺ thus suggesting that MPP⁺ toxicity was mediated by the activation of p38. Prior to this experiment, a dose response was performed to standardize the optimum dosage concentration for p38 inhibitor (SB239063). SB239063 conferred complete protection at 1 μM concentration against toxicity caused by 10 μM MPP⁺ (Fig. 35) and did not show further protection at higher concentration i.e. [5 μM (Fig. 36A, B)] and was toxic to these cells at further higher concentration [10 μM (Fig. 36A, B)].

MPP⁺ mediates cytotoxicity via nuclear translocation of p53 which is prevented by SB239063: p53 is normally localized in the cytoplasm of primary neurons derived from the human CNS progenitor cells. In response to an apoptotic stimuli, p38 is known to phosphorylate p53 triggering its nuclear translocation leading to activation of target genes. Exposure of cells to MPP⁺ (10 μM) for 24 h, induced translocation of p53 to the nucleus which was prevented by pretreating the cells with p38 inhibitor, SB239063 (Fig. 37), indicating the role of p38 in the translocation of p53.

Figure 35: p38 inhibitor (SB239063) protects primary neurons derived from human CNS progenitor cells from MPP$^+$ mediated cytotoxicity: Cells were treated with either vehicle or SB239063 (1 µM) for 60 min before exposure to MPP$^+$ (10 µM) for 24 h and subjected to TUNEL assay. SB239063 conferred protection against MPP$^+$ cytotoxicity. Quantification represents percentage of cells positive for TUNEL staining per total number of cells as represented by DAPI staining. Data is represented as mean ±SD of four independent experiments. Asterisks indicate values significantly different from controls ($p<0.05$).

Figure 36: Higher dose concentration of SB239063 does not protect cells from MPP$^+$ mediated cytotoxicity, rather it itself causes toxicity.

Primary neurons derived from human CNS progenitor cells were treated with either vehicle or SB239063 (5/10 µM) for 60 min before exposure to MPP$^+$ (10 µM) for 24 h and subjected to TUNEL assay. SB239063 at 10 µM caused toxicity and did not protect cells against MPP$^+$ cytotoxicity whereas at 5 µM concentration it afforded protection as was observed with 1 µM. Quantification represents percentage of cells positive for TUNEL staining per total number of cells as represented by DAPI staining. Data is represented as mean ±SD of four independent experiments. Asterisks indicate values significantly different from controls ($p<0.05$).

Figure 37: p38 inhibitor SB239063 protected cells against MPP$^+$ cytotoxicity by preventing p53 translocation into the nucleus.

Primary neurons derived from human CNS progenitor cells were pretreated with vehicle or SB239063 (1 µM) for 60 min before exposure to MPP$^+$ (10 µM) for 24 h and immunostained for p53 to study its subcellular localization. MPP$^+$ induced nuclear translocation of p53 was inhibited by SB239063 thereby preventing cytotoxicity. Total number of cells is represented by propidium iodide (PI) staining. Data are represented as mean ± SD of four independent experiments. Asterisks indicate values significantly different from controls ($p<0.05$).

MPP⁺ mediated cell death is prevented by pifithrin-α and JNK inhibitor (SP600125) but SP600125 partially prevent the nuclear translocation of p53: To discern if MPP⁺ induced translocation of p53 is mediated via activation of JNK and whether cell death caused by MPP⁺ is also mediated via p53 activation, we pretreated cells with pifithrin-α or JNK inhibitor (SP600125) for 60 min and further treated them with MPP⁺ for 24 hr. Cytoxicity caused by MPP⁺ was prevented by both pifithrin-α and SP600125 (Fig. 38A & B). However, unlike to what was observed with SB239063, pretreatment of cells with SP600125 attenuated the nuclear translocation of p53 minimally (Fig. 39) but did not abolish it suggesting that phosphorylation of p53 was independent of JNK activation.

Antioxidant α-lipoic acid attenuates but does not abolish MPP⁺ induced cytotoxicity: MPP⁺ also induces generation of ROS which can be quenched by an antioxidant such as α-lipoic acid and subsequent cell death could be prevented. Primary dopaminergic cultures derived from human CNS progenitor cells were treated with α-lipoic acid (100 μM) for 60 min prior to the treatment with MPP⁺ for 24 hr. Cell death was estimated by TUNELassay. Cytotoxicity induced by MPP⁺ is only partially attenuated by treatment with α-lipoic acid (Fig. 40) suggesting that besides generation of ROS and subsequent mitochondrial dysfunction, MPP⁺ triggered other apoptotic cell death mechanisms which could not be completely prevented by α-lipoic acid at the dose used in this experiment.

Figure 38: JNK inhibitor (SP600125) and pifithrin-α confers protection against MPP⁺ mediated cell death.

Cells were treated with either (A) JNK inhibitor (SP600125; 5 µM) or (B) pifithrin-α (250 nM) for 60 min prior to exposure to MPP⁺ (10 µM) for 24 h. Controls were treated with vehicle. Both pifithrin-α and SP600125 completely prevented MPP⁺ mediated cell death. Scale bar represents 200 µm. Quantification represents, percentage of cells positive for TUNEL staining per total number of cells as represented by DAPI count.

Figure 39: JNK inhibitor (SP600125) although protects against MPP⁺ mediated cell death but only attenuates nuclear translocation of p53 partially.

p53 is normally localized in the cytoplasm in primary neurons under physiological conditions and gets translocated into the nucleus in response to MPP$^+$ exposure. The nuclear translocation of p53 was partially prevented by pretreating the cells with JNK inhibitor, SP600125. Total number of cells is represented by propidium iodide (PI) staining. Data are represented as mean ± SD of four independent experiments. Asterisks indicate values significantly different from controls ($p<0.05$).

Figure 40: Antioxidant α-lipoic acid does not abolish but attenuates MPP⁺ induced cytotoxicity.

Cells were treated with either vehicle or α-lipoic acid (100 µM) for 60 min prior to further treatment with MPP⁺ (10 µM) for 24 h. α-lipoic acid partially attenuated MPP⁺ induced cell death. Scale bar represents 200 µm. Quantification represents, percentage of cells positive for TUNEL staining per total number of cells as represented by DAPI count.

Daxx is translocated from the nucleus to the cytoplasm in response to MPP$^+$ in SHSY-5Y neuroblastoma cells: Daxx, the death-associated protein is a transcriptional repressor that is normally present in the nucleus. Daxx is known to translocate to cytoplasm, associate with ASK1, and participate in the apoptotic process (Khelifi et al., 2005). To investigate if cell death mediated by MPP$^+$ is Daxx dependent and to study the subcellular localization of Daxx in response to MPP$^+$, SHSY-5Y cells were treated with varying concentration of MPP$^+$ for 24 hr to standardize the optimum dose and cell viability was studied using MTT assay (Fig. 41A), control were treated with vehicle for the same time. MPP$^+$ at 3 mM caused more than 50% cell death and was found to be appropriate for this cell line. To standardize the optimum time point at which 3 mM MPP$^+$ induces cytosolic translocation of Daxx, SHSY-5Y cells were treated with 3 mM MPP$^+$ for 4, 8 and 12 hr and subcellular localization of Daxx was examined by immunostaining. Daxx was found to be completely translocated after 12 hr thus preceding cell death which was observed at 24 hr (Fig. 41B).

Cytosolic translocation of Daxx caused by MPP$^+$ is prevented by pretreatment of SHSY-5Y with anti-oxidants: Cytosolic translocation of Daxx caused by treatment with 3 mM MPP$^+$ for 12 h was prevented by treating cells with N-acetyl cysteine (100 μM) or ALA (100 μM) for 1 h prior to MPP$^+$ treatment (Fig. 42).

Figure 41: Dose and time response of MPP⁺ in SHSY-5Y cells for cell viability and translocation of Daxx.

(A) SHSY-5Y cells were treated with either vehicle or MPP⁺ at varying concentration for 24 hr and cell viability was assessed using MTT assay. (B) Cells were treated with 3 mM MPP⁺ for 4, 8 and 12 hr and translocation of Daxx was examined by immunostaining. Controls were treated with vehicle for 12 hr. Bar represents 50 μm.

Figure 42: Daxx translocates from nucleus to the cytosol following exposure to MPP_ in SH-SY5Y human neuroblastoma cells.

Cells were exposed for 12 h to MPP$^+$ alone (3 mM) or in combination with N-acetyl cysteine (NAC; 100 µM) or ALA (ALA; 100 µM) and then immunostained for Daxx. Cells were counterstained with DAPI prior to mounting. On exposure to MPP$^+$ Daxx translocated to cytosol, which was prevented by prior treatment with NAC or ALA. Scale bar 120 µm.

1) Redox driven cell death signaling cascade activated in males are restrained in females.

Incidence of Parkinson's disease (PD) has been reported to be lower in women as compared to men (1:1.46 respectively; (Taylor et al., 2007). Although neuroprotective effects of estrogen are well acknowledged, the underlying mechanisms are poorly understood. To understand the mechanism of neuroprotection offered by estrogen, a comparative evaluation of events was performed in male and female mice, in response to MPTP administration. The sex difference seen in human population is reflected in the animal model wherein female mice are protected from MPTP toxicity (Morissette et al., 2008). We examined the molecular mechanisms underlying the neuroprotection seen in females by examining the redox activated MAP kinase signaling cascades that have been implicated in the selective degeneration of SNpc dopaminergic neurons in mice (Karunakaran et al., 2007b; Karunakaran et al., 2008).

MPTP induced loss of dopaminergic neurons is attenuated in female mice: We examined the loss of tyrosine hydroxylase positive neurons in SNpc following administration of MPTP for 8 and 14 days (Fig. 43A). MPTP administration caused 21% loss of dopaminergic neurons in male mice as compared to 12.5% loss in female mice after 8 days while the corresponding loss was 47% and 22.5% after 14 days of treatment (Fig. 43B). When we looked at the cell loss in SNpc after 8 days, we found that the difference between males and females was small (21% vs 12.5%), albeit significant. However, between 8 and 14 days, while the number of neurons declined sharply in males (47%), the females showed less decline (22.5%; Fig. 43C). The number of TH positive neurons in the saline treated groups in male and female SNpc was not significantly different. They were 1885 ± 96 and 1918 ± 28 neurons respectively, in males and females. We validated the above by counting Nissl positive neurons in SNpc and observed that

MPTP treatment for 8 days caused 18% and 12.5% loss in male and female mice, respectively (Fig. 43D). Although the cell loss was only attenuated but not eliminated in female mice, the mitochondrial dysfunction seen as inhibition of complex I activity was completely abolished in female mice after 8 days of MPTP treatment (Fig. 43E). This indicates that in addition to mitochondrial complex I dysfunction, other cell death pathways may operate in MPTP mediated degeneration.

Activation of apoptotic signal-regulating kinase I (ASK1) does not occur in female mice treated with MPTP: ASK1, a MAPKKK is activated in the ventral midbrain after MPTP treatment in male but not female mice where the ASK1 activation was found to be significantly reduced (Fig. 44A). Under normal conditions, ASK1 is bound to reduced thioredoxin (Trx), a protein disulfide oxidoreductase preventing its autophosphorylation at serine 845. Oxidation of the cysteine thiols in Trx results in its dissociation from ASK1 triggering its autophosphorylation, which in turn propagates the downstream death cascade (Saitoh et al., 1998). Thus, the levels of thioredoxin and thioredoxin reductase are critical determinants of ASK1 activation (Fig. 44B).

Figure 43: Loss of dopaminergic neurons and inhibition of complex I activity is more pronounced in MPTP treated male mice than female.
Animals were treated with a single dose of vehicle or MPTP daily, for 8 or 14 days following which they were sacrificed on 9th or 15th day. (A) Tyrosine hydroxylase (TH) immunostaining in male and female mice treated with vehicle (saline; control), and MPTP for 8 days & 14 days. Bar represents 50 μm. (B) Quantitative stereological analysis of tyrosine hydroxylase (TH) positive neurons in male and female mice treated with vehicle or MPTP for 8 or 14 days, showing significant loss of TH positive neurons in males, after 14 days of MPTP administration. Values are mean ± SD (n=3). (C) The data was analyzed to determine the rate of degeneration, and is represented as a solid line (_____) for male mice, while for females it is depicted as a dotted line (......). (D) Quantitative stereological analysis of Nissl stained cells present in SNpc, in male and female mice treated with vehicle or MPTP for 8 days. Values are mean ± SD (n=3). (E) Complex I activity is inhibited in midbrain & striatum of male mice treated with MPTP for 8 days whereas females and vehicle treated controls were unaffected. Activity is expressed as nmoles of NADH oxidized/min/mg protein. Values are mean ± SD (n=3). Asterisks indicate values significantly different from vehicle treated controls.

Figure 44: Activation of apoptotic signal regulating kinase 1 (ASK1) is suppressed in females but not in male mice following MPTP administration.
Animals were treated with a single dose of MPTP or vehicle and sacrificed 12 hr later. (A) Representative immunoblot from midbrain of male and female mice treated with vehicle (Control, C) and MPTP (12 hr) depicting the protein levels of phosphorylated-ASK1 (Thr 845) and ASK1. β-tubulin levels were measured as loading control. Densitometric analysis of the immunoblots representing the relative intensity of the immunoreactive bands, showing upregulation of pASK only in MPTP treated males. Values are mean ± SD (n=6). Asterisks indicate values significantly different from vehicle treated controls. (B) Schematic representation of mechanism of autophosphorylation of ASK1 is depicted. ROS generated during oxidative stress causes oxidative modification of Trx, which in its reduced state remains bound to ASK1 but dissociates once it is oxidized thus releasing ASK1 and triggering its autophosphorylation. Higher levels of reduced Trx prevent autophosphorylation of ASK1.

Enhanced expression of thioredoxin, thioredoxin reductase and glutathione reductase in midbrain and striatum of female mice: Higher expression of thioredoxin measured as total protein levels was seen in midbrain and striatum of female mice compared to males (Fig. 45A). Thioredoxin reductase is responsible for maintaining thioredoxin in its reduced form. Higher constitutive expression of thioredoxin reductase (TR; Fig. 45B) was also observed in the midbrain and striatum of female mice. Moreover, the activity of glutathione reductase (GR), a key enzyme involved in reduction of GSSG to GSH, was found to be higher in the whole brain; both in postmitochondrial supernatant (PMS) and mitochondrial pellet (Mito; Fig. 46A) as well as in the midbrain and striatum of female mice (Fig. 46B). Higher expression of glutathione reductase was also measured as total protein levels in midbrain and striatum of female mice compared to males (Fig. 46C).

Downstream of ASK1 activation, phosphorylation of p38 MAP kinase occurs in males but not female mice SNpc: Activation of ASK1 leads to phosphorylation of downstream targets, such as p38 MAP kinase. Increased phosphorylation of p38 was observed 12 hr following MPTP treatment only in ventral midbrains of male mice (Fig. 47A) whereas no such activation of p38 was observed in midbrains of female mice. Immunohistochemical analysis co-localizing pp38 and TH clearly demonstrated the increased phosphorylation of p38 in TH positive neurons in the SNpc of male mice (Fig. 47B), which is clearly discernible at low and high magnification. In female mice, p38 activation was not observed in the TH positive neurons in SNpc (Fig. 47C).

Figure 45: Higher constitutive expression of thioredoxin and thioredoxin reductase in midbrain and striatum of female mice.

Constitutive expression of enzymes involved in thioredoxin system was quantitated by immunoblotting. Representative immunoblots from midbrain and striatum of male (M) and female (F) animals depicting the protein levels of (A) thioredoxin and (B) thioredoxin reductase. β-tubulin levels were measured as loading control. Densitometric analysis of the immunoblots representing the relative intensity of the immunoreactive bands, showing higher expression of (A) thioredoxin and (B) thioredoxin reductase in females. Values are mean ± SD (n=3 animals).

Figure 46: Higher constitutive expression of glutathione reductase in midbrain and striatum of female mice.

Activity and constitutive expression of glutathione reductase was quantitated by enzymatic assay and immunoblotting. (A) Glutathione reductase activity measured in whole brain postmitochondrial supernatant (PMS) and mitochondrial pellet (Mito) was found to be higher in females. (B) Activity was also found to be higher in females when measured in striatum (ST) and midbrain (MB). Activity is expressed as nmoles of NADPH oxidized/min/mg protein. Values are mean ± SD (n=3). (C) Representative immunoblots from midbrain (MB) and striatum (ST) of male (M) and female (F) animals depicting the protein levels of glutathione reductase. β-tubulin levels were measured as loading control. Densitometric analysis of the immunoblots representing the relative intensity of the immunoreactive bands, showing higher expression of glutathione reductase in females. Values are mean ± SD (n=3 animals). Asterisks indicate values significantly different from vehicle treated controls.

Figure 47: Activation of p38 in male but not in female mice after MPTP administration.

Animals were treated with a single dose of MPTP or vehicle and sacrificed 12 or 24 hr later. (A) Representative immunoblot from midbrain of male and female mice treated with vehicle (control; C) and MPTP (12 hr) depicting the protein levels of phosphorylated p38 and p38. β-tubulin levels were measured as loading control. Densitometric analysis of the immunoblots representing the relative intensity of the immunoreactive bands are shown below, higher levels of phosphorylated-p38 found in MPTP treated males only. Values are mean ± SD (n=7). Asterisks indicate values significantly different from vehicle treated controls. (B) Immunohistochemical co-localization revealed the presence of phosphorylated p38 (pp38; FITC) in the soma of tyrosine hydroxylase positive (TH; Texas red) neurons of SNpc, in the ventral midbrain of male mice, after 24 hr of exposure to MPTP, whereas control animals show low levels of pp38. Bar represents 200 μm (B; upper panel). Lower panel shows corresponding magnified images of the neurons in upper panel. Bar represents 20 μm (B; lower panel). (C) Co-immunostaining for pp38 and tyrosine hydroxylase revealed negligible presence of pp38 in the soma of tyrosine hydroxylase positive neurons of SNpc and ventral midbrain of female mice 24 hr after exposure to MPTP or vehicle. Bar represents 200 μm (C; upper panel). Lower panel shows corresponding magnified images of the neurons in upper panel. Bar represents 20 μm (C; lower panel).

Loss of nuclear DJ-1 and translocation of Daxx to the cytosol was not seen in females: DJ-1, a redox sensing protein sequesters Daxx (a transcriptional repressor and death associated protein) within the nucleus and prevents its association with ASK1 and subsequent propagation of death cascade (Junn et al., 2005). MPTP exposure led to significant decrease in nuclear DJ-1 levels in the ventral midbrain of male mice, whereas females were unaffected (Fig. 48A). Further, the constitutive expression of DJ-1 was higher in the ventral midbrain of female mice indicating its possible regulation by estrogen (Fig. 48B). In female mice, translocation of Daxx from the nucleus to the cytosol also did not occur either in SNpc or VTA neurons, unlike in males where Daxx clearly translocated in SNpc neurons but not in VTA (Fig. 48C and D). Thus, decrease in nuclear DJ-1 levels triggers cytosolic translocation of Daxx which could potentially associate with ASK1 in the cytoplasm and propagate cell death cascade (Fig. 48E).

Figure 48: Reduced nuclear DJ-1 levels and cytosolic translocation of Daxx in male but not female mice treated with MPTP.
Animals were treated with a single dose of MPTP or vehicle and sacrificed 12 or 24 hr later. (A) Representative immunoblot for nuclear DJ-1 from ventral midbrain of male and female mice treated with saline (S) and MPTP (12 hr) is depicted. Densitometric analysis of the immunoblots representing the relative intensity of the immunoreactive bands shows significant decrease in DJ-1 levels in MPTP treated males. (B) Representative immunoblots for extranuclear DJ-1 from ventral midbrain of male and female mice. Densitometric analysis of the immunoblots show higher constitutive expression of DJ-1 in females. Values are mean $\pm$ SD (n=7). Asterisks indicate values significantly different from vehicle treated controls. (C) Immunohistochemical localization of Daxx in SNpc, revealed its translocation to the cytosol in male mice after 24 hr of MPTP administration whereas it was retained within the nucleus in female mice and controls. Bar represents 50 μm (C; upper panel). Lower panel show corresponding magnified images of the neurons in upper panel. Bar represents 10 μm (C; lower panel). (D) Immunohistochemical localization of Daxx in VTA reveals its retention within the nucleus after MPTP administration in both male and female mice. Bar represents 50 μm (D; upper panel). Lower panel show corresponding magnified images of the neurons in upper panel. Bar represents 10 μm (D; lower panel). (E) Schematic representation showing nuclear translocation of Daxx triggered by decreased nuclear DJ-1 levels following which Daxx associates with ASK1 in the cytosol and propagates the death cascade.

DISCUSSION

Neurodegenerative disorders, including PD have been shown to involve oxidative stress, mitochondrial dysfunction and dysregulation of protein turnover machinery as mechanisms of their pathogenesis. Oxidative stress, a hallmark for familial as well as sporadic form of PD is caused by excessive generation and accumulation of reactive oxygen/nitrogen species (ROS/RNS), which cannot be efficiently neutralized by cell's antioxidant system. ROS generated during oxidative stress causes lipid peroxidation, damage to DNA and oxidative modification of several redox sensitive proteins leading to the change in their conformation and altered biological activity. Alteration in oxidation state of these proteins in turn regulates the redox signaling pathways. Normally, ROS are signal regulators in metabolic processes and cell normally harbors multiple antioxidant defense systems to keep a check on the physiological generation of these reactive species. However, their uncontrolled generation and accumulation causes reversible or irreversible oxidative modifications of protein thiol groups thus adversely affecting their function and thereby redox regulatory mechanism.

Maintenance of protein thiol homeostasis is critical for the normal physiological functions of proteins and may be perturbed in presence of oxidative insult causing redox dysregulation. Different antioxidant systems within a cell help in regulation of the redox environment, glutathione (GSH) being the most abundant cellular thiol. Besides GSH, thiol disulfide oxidoreductases (TDORs) such as glutaredoxin and thioredoxin are potent antioxidant systems, responsible for modulating protein function by acting on thiol disulfides (Kalinina et al., 2008, Gravina and Mieyal, 1993). Development of the disease is accompanied by gradual decrease in glutathione content and increased oxidative stress which induces neuronal apoptosis involved in degenerative process of dopaminergic neurons (Chinta and Andersen, 2008; Ballatori et al., 2009).

Previous data from our laboratory suggests that Grx1 protects mitochondrial complex I against toxicity induced by 1-methyl-4-phenyl-1,2,3,6-tetrahydropyridine (MPTP) and β-N-oxalylamino-L-alanine (L-BOAA), in mouse model for PD and motor neuron disease respectively (Kenchappa et al., 2002; Kenchappa and Ravindranath, 2003; Kenchappa et al., 2004; Diwakar et al., 2007). Further, downregulation of Grx1 by antisense oligonucleotides resulted in loss of complex I activity, which could be fully reversed back by DTT, thereby indicating that Grx1 prevented oxidative modification of thiol groups present in complex I subunits. Grx1 downregulation also exacerbated the extent of complex I inhibition caused by both MPTP and L-BOAA in mice, thus emphasizing on the significance of Grx1 in maintenance of function of mitochondrial complex I. However, the fact that Grx1 being a cytosolic enzyme modulates the function of complex I was questionable and further study was warranted to understand its protective effects on mitochondrial function. First part of the current work unveils the mechanism of how Grx1, being a predominantly cytosolic enzyme modulates the mitochondrial function.

PD is a chronic progressive degenerative disorder characterized by selective loss of mesencephalic dopaminergic neurons residing in substantia nigra pars compacta (SNpc) in the ventral midbrain. It is primarily sporadic in incidence, oxidative stress being the main causative factor. Familial component of the disease is less prevalent and characterized by loss or gain of function mutations in several genes. DJ-1 is a redox sensitive stress responsive protein and its loss of function mutations are implicated in PD. During oxidative stress, it scavenges ROS and in turn undergoes oxidative modification at the cysteine residues present at position 46, 53 and 106. Reversible oxidative modification of the cysteine residue at 106 position triggers it to translocate to mitochondria, which has been reported as a protective mechanism. To function as a ROS scavenger, DJ-1 should be maintained in its reduced form and should not have undergone irreversible oxidation. Familial PD is associated with loss of function of DJ-1 due to mutations in it, but functional impairment due to its irreversible oxidation could be anticipated in sporadic PD, since the extent of oxidative stress is high during the course of the disease. The second and major part of the research illustrates work done to investigate if perturbation of redox homeostasis results in loss of function of DJ-1, since it may have implications in the pathogenesis in sporadic PD.

Earlier data from our lab demonstrates that although MPTP induced complex I inhibition could be restored completely by administration of α-lipoic acid (ALA) to mice but cell death of SNpc neurons per se could not be completely prevented. Further, data from our lab showing activation of apoptotic signal regulating kinase1 (ASK1; MAPKKK) in response to MPTP indicated that apart from mitochondrial dysfunction, oxidative stress induced by MPTP could potentially perturb redox regulated signaling pathways resulting in the activation of redox driven cell death cascade which could cause the selective demise of dopaminergic neurons in SNpc. In order to decipher if aberrant cells death pathways were elicited selectively in dopaminergic neurons of SNpc, MAP kinase pathway downstream to ASK1 was analyzed in dopaminergic neuronal cultures of human fetal progenitor cells, using MPP^+ as the model neurotoxin.

There are mounting epidemiological evidences establishing the fact that incidence and prevalence of PD is more in men when compared to women and the same is reflected in animal models of PD. Data from our lab also suggests that MPTP induced complex I inhibition observed in male mice does not occur in females, however pretreatment of female mice with ICI 182,780, estrogen receptor (ER) antagonist sensitizes them to MPTP-mediated complex I dysfunction (Kenchappa et al., 2004). Although, the fact that estrogen affords neuroprotection is well acknowledged, the mechanisms underlying estrogen mediated neuroprotection are not very clear. Given that, redox regulated cell death mechanisms are activated in MPTP administered male mice and constitutive expression of Grx1, a redox regulating enzyme is reported to be more in female, differential regulation of redox driven apoptotic pathways in male and female mice in response to MPTP was worth exploring. With this rationale, differential response in male and female mice was studied in terms of activation of apoptotic cascade after MPTP administration.

I. Role of Glutaredoxins in Neuroprotection

4) Knockdown of cytosolic glutaredoxin 1 leads to loss of mitochondrial membrane potential: Implication in neurodegenerative diseases.

Mitochondria are both the power centers of the cell as well as mediators of cell death through apoptosis and they have been implicated in a variety of neurodegenerative disorders. Although mechanisms underlying mitochondrial dysfunction leading to neurodegeneration are not entirely clear, it is generally believed that oxidative stress is a key player in some of these events (Beal, 2007). Thiol disulfide oxido-reductases, such as glutaredoxin and thioredoxin, are a group of enzymes that maintain redox homeostasis during oxidative conditions. Glutaredoxin 1 senses cellular redox potential and catalyzes glutathionylation, an important redox regulatory mechanism in response to oxidative stress (Fernandes and Holmgren, 2004).

During oxidative stress, GSH is oxidized to GSSG (Eqn. 1), which is often effluxed out of the cell, thereby, preventing the oxidative modification of protein thiols (PrSH) to protein glutathione mixed disulfide (PrSSG; Eqn. 2). Protein thiols are also oxidized by ROS in a sequential manner to sulfenic, sulfinic and sulfonic acid (Eqn. 3). Sulfenic acids further react with GSH to form PrSSG thus preventing their irreversible oxidation to sulfonic acids (Eqn. 4). PrSSG may be further modified to generate intramolecular and intermolecular protein mixed disulfides (PrSSPr; Eqn. 5). PrSSGs can be reduced back to protein thiols effectively by Grx1 (in the cytosol) and Grx2 (in the mitochondria) utilizing GSH and reducing equivalents of NADPH (Eqn. 6). The GSSG generated by this reaction is reduced back to GSH by glutathione reductase using reducing equivalents from NADPH (Eqn. 7). Downregulation of Grx1 would therefore potentially lead to increased levels of protein glutathione mixed disulfides as also mixed disulfides of proteins (PrSSPr).

$$2GSH \xrightarrow{ROS} GSSG \qquad \qquad ....(1)$$

$$PrSH + GSSG \longrightarrow PrSSG + GSH \qquad \qquad ....(2)$$

$$PrSH \xrightarrow{ROS} PrSOH \xrightarrow{ROS} PrSO_2H \xrightarrow{ROS} PrSO_3H \qquad ....(3)$$

$$PrSOH + GSH \longrightarrow PrSSG \qquad \qquad ....(4)$$

$$Pr\genfrac{}{}{0pt}{}{S-SG}{SH} \longrightarrow Pr\genfrac{}{}{0pt}{}{S}{S} + GSH \qquad \qquad ....(5)$$

$$PrSSG \xrightarrow{Grx/GSH/NADPH} PrSH + GSSG \qquad \qquad ....(6)$$

$$GSSG \xrightarrow{GR/NADPH} 2GSH \qquad \qquad ....(7)$$

Grx1 is a cytosolic protein and therefore its ability to influence mitochondrial function in this manner is not anticipated. In this study, we examined the role of Grx1 in maintenance of mitochondrial membrane potential by downregulating Grx1 using shRNA in neuroblastoma cells. Further, we also exposed cells to the excitatory amino acid L-

BOAA and studied the interplay between Grx1 and mitochondrial membrane potential. Previous data from our lab has demonstrated that constitutive expression of Grx1 is significantly higher in female brain regions as compared to the male brain and downregulation of Grx1 in female mouse brain renders them vulnerable to L-BOAA mediated mitochondrial dysfunction in a manner similar to that seen in male mice (Diwakar et al., 2007). We therefore also studied the effects of estrogen on expression of thiol disulfide oxidoreductases, particularly Grx1, and ability of estrogen to protect against mitochondrial dysfunction mediated by L-BOAA.

Grx1 was predominantly localized in cytosol as opposed to the presence of Grx2 in mitochondria and observed that downregulation of Grx1 using shRNA in Neuro-2a cells resulted in generation of ROS and increase in levels of free cytosolic calcium ion concentration. Further probing into the mechanism of ROS mediated mitochondrial dysfunction revealed, that downregulation of Grx1 dramatically altered the mitochondrial membrane potential (MMP) and the loss of MMP was proportional to the decrease in levels of Grx1 following its knockdown. Optimum decrease of Grx1 was found 72 hr after transfection which correlated with the maximum loss of MMP. This loss of MMP could be restored by α-lipoic acid and cyclosporine A as demonstrated by quantitative determination of change in MMP using TMRM as the indicator dye. Both, α-Lipoic acid, an endogenous thiol antioxidant and cyclosporine A, an inhibitor of peptidyl prolyl-cis, trans-isomearse (PPIase) activity which in turn blocks mitochondrial permeability transition pore, (Galat and Metcalfe, 1995), effectively prevented MMP loss mediated by Grx1 downregulation. While cyclosporine A is known to inhibit Ca^{2+} induced mitochondrial permeability transition (Mirzayan et al., 2008; Devinney et al., 2009), the ability of α-lipoic acid to afford protection against the MMP loss could be due to the fact that the oxidative modification of protein thiols was restricted to the formation of sulfinic acid and mixed disulfides which can be reversed enzymatically by sulfiredoxins and glutaredoxin, respectively (Findlay et al., 2006; Gallogly and Mieyal, 2007; Jonsson and Lowther, 2007; Rhee et al., 2007). However, if not disrupted, oxidative modification of protein thiols would proceed further in a stepwise manner leading to the formation of sulfinic acid(s) (Salmeen et al., 2003) and sulfonic acid derivatives which would be irreversible (Fig. D1). This effect of downregulation of Grx1 was noted in two neuronal cell lines, SH-SY5Y and Neuro-2a of human and mouse origin respectively.

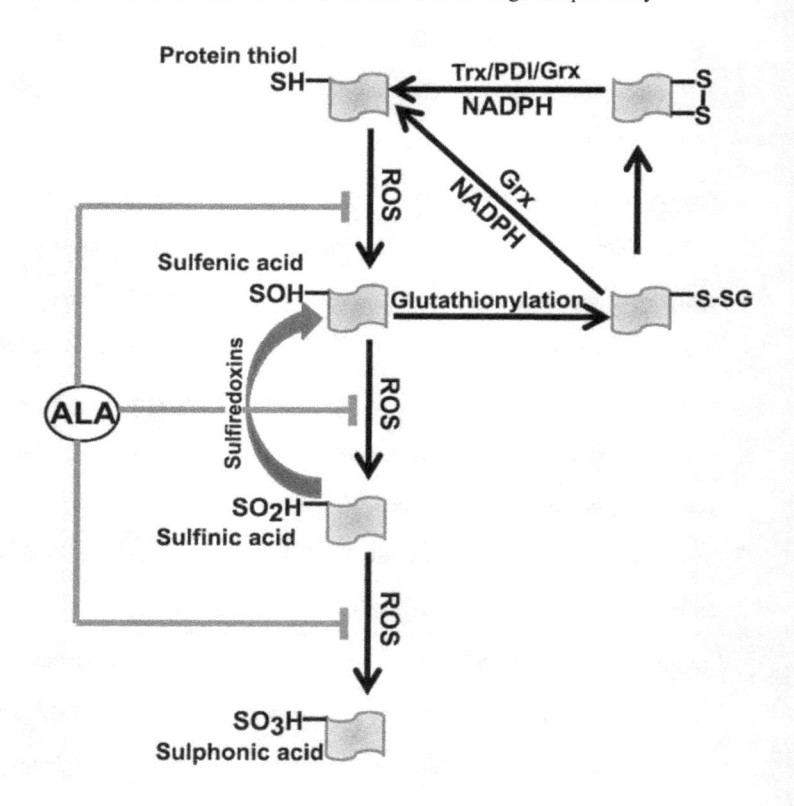

Figure D1: Oxidative modification of thiol groups in proteins.
Protein thiols may be oxidized sequentially to sulfenic, sulfinic and sulfonic acid. Sulfenic acids can react with GSH to form PrSSG thus preventing their irreversible oxidation to sulfonic acids. PrSSG may be further modified to protein mixed disulfides. PrSSG are reduced back to protein

thiols very effectively by Grx1 utilizing GSH and reducing equivalents of NADPH. The thiol antioxidant, α-lipoic acid, can potentially prevent the oxidative modification of protein thiols.

Next, we studied the role of Grx1 in maintenance of mitochondrial potential by performing rescue experiments using L-BOAA as an oxidative stimulus. L-BOAA is an excitatory amino acid that causes mitochondrial dysfunction (Diwakar and Ravindranath, 2007). Exposure of SH-SY5Y cells to L-BOAA resulted in loss of MMP and cell death, however, in cells overexpressing Grx1, L-BOAA was unable to alter MMP adversely or affect cell viability. Loss of MMP caused by L-BOAA could also be prevented by cyclosporine A, similar to what was observed following Grx1 knockdown, thereby suggesting that the mechanism underlying mitochondrial permeability transition in both the paradigms was similar.

Thiol disulfide oxidoreductases like glutaredoxin and thioredoxin are regulated by estrogens as reported earlier (Sahlin et al., 1997; Kenchappa et al., 2004). We investigated if redox regulating enzymes of glutaredoxin (Grx) and thioredoxin (Trx) system were in turn regulated by estrogen in SH-SY5Y cells. The rationale for using SH-SY5Y was that it expresses both estrogen receptors α and β, and we could detect this phenomenon in these cells. Expression of several enzymes of Grx and Trx sytem such as Grx1, Trx1, Trx2 and TR was induced by 17-β estradiol and this induction could be prevented by estrogen receptor antagonist, ICI 182,780, thereby indicating their regulation by estrogen. Expression of Grx1 is regulated by estrogen in the central nervous system (CNS) and ovariectomy downregulates Grx1 levels in brain regions as has been reported from our laboratory earlier (Kenchappa et al., 2004). Exposure of SH-SY5Y cells to estrogen upregulated Grx1 and prevented L-BOAA mediated loss of MMP clearly delineating the important role of Grx1 in maintaining mitochondrial function and thereby the cell viability.

MMP is regulated by critical proteins such as VDAC, ANT and cyclophilin D, localized in outer mitochondrial membrane, inner mitochondrial membrane and mitochondrial matrix respectively. The redox status of these proteins is known to affect MMP. Evidences suggest that the oxidative state of the vicinal thiol groups in VDAC are known to critically affect MMP and thereby, the opening of mitochondrial permeability transition pore (mPTP) (Petronilli et al., 1994). Similarly, the redox status of ANT is also known to be crucial for the maintenance of MMP (Haouzi et al., 2002; Halestrap and Brennerb, 2003; Yang et al., 2007). This encouraged us to examine the redox status of both VDAC and ANT following downregulation of Grx1 and we found that while there was a dramatic loss of reduced VDAC, reduced ANT did not show a similar loss. Interestingly there was an increase in the amount of reduced ANT measured both as the derivatized and the reduced protein (Fig. 16). This could be due to the fact that ANT is unaltered by downregulation of Grx1 in cytosol and insufficient downregulation of Grx1 present in IMS of mitochondria. Other than this, there is a possibility that the mitochondrial glutaredoxin, Grx2, efficiently reduces the oxidized ANT. Although the expression of Grx2, the mitochondrial form of glutaredoxin located predominantly in the matrix, was unchanged when Grx1 was downregulated (Fig. 2B), its potential role in maintaining the reduced state of key mitochondrial proteins cannot be ruled out.

Mieyal and colleagues have demonstrated the presence of Grx1 in the intermembrane space in the mitochondria (Pai et al., 2007). The knockdown of Grx1 should potentially downregulate the expression of Grx1 both in the cytosol and the intermembrane space of the mitochondria, although it is presumable that the turnover rate of Grx1 could be different in the cytosol and the intermembrane space of the mitochondria and therefore knockdown of Grx1 may not be similar in the two subcellular compartments. This may, to an extent, explain the lack of oxidative modification of the cysteine residue(s) in ANT. The role of Grx1 in redox regulation in the intermembrane space is yet to be clearly understood and additional studies are considered necessary in this direction.

Results obtained from this study indicate that the oxidative modification of VDAC by downregulation of Grx1 could be a key player in mediating the loss of MMP. Several isoforms of VDAC are known to exist, some of which may not have any effect on MMP. In a study, mitochondria isolated from VDAC1 deficient cells underwent mitochondrial permeability transition normally, although it could have resulted due to the compensation by other isoforms of VDAC, such as VDAC2 and VDAC3 (Baines et al., 2007). However, others have provided evidence for the importance of VDAC in maintenance of MMP (Kim et al., 2005a; Kim et al., 2006; Yagoda et al., 2007). Several findings point towards the importance of VDAC in mitochondrial permeability transition, such as the electrophysiological properties of permeability transition pore which shows a striking similarity to those of VDAC (Szabo et al., 1993; Szabo and Zoratti, 1993). Further, factors such as Ca^{2+}, NADH and glutamate that alter VDAC channel properties, also modulate permeability transition pore (Costantini et al., 1996; Fontaine et al., 1998). VDAC in association with ANT, has also been purified by chromatography of mitochondrial extracts on a cyclophilin D affinity column (Crompton et al., 1998).

In the current study, it was demonstrated for the first time that perturbation of the redox milieu in the cytosol through knockdown of Grx1 could modify VDAC, which could potentially result in the subsequent loss of MMP. Our results also specify that even though the reduced form of ANT is not decreased, oxidative modification of VDAC could lead to loss of MMP. In view of the fact that the reduced form of the inner membrane protein ANT did not decrease by knockdown of Grx1, we did not monitor the redox status of cyclophilin D, which resides in the mitochondrial matrix. Significance of this study is underlined by the fact that downregulation of the cytosolic Grx1 can lead to mitochondrial dysfunction and this could be mediated by loss of MMP, which occurs through redox modification of VDAC but not ANT in the cell lines examined (Fig. D2). Hence, maintenance of protein thiol homeostasis through this vital oxidoreductase, Grx1, may play a significant role in disease pathogenesis such as Parkinson's disease and motor neuron disease, wherein mitochondrial dysfunction is a key player.

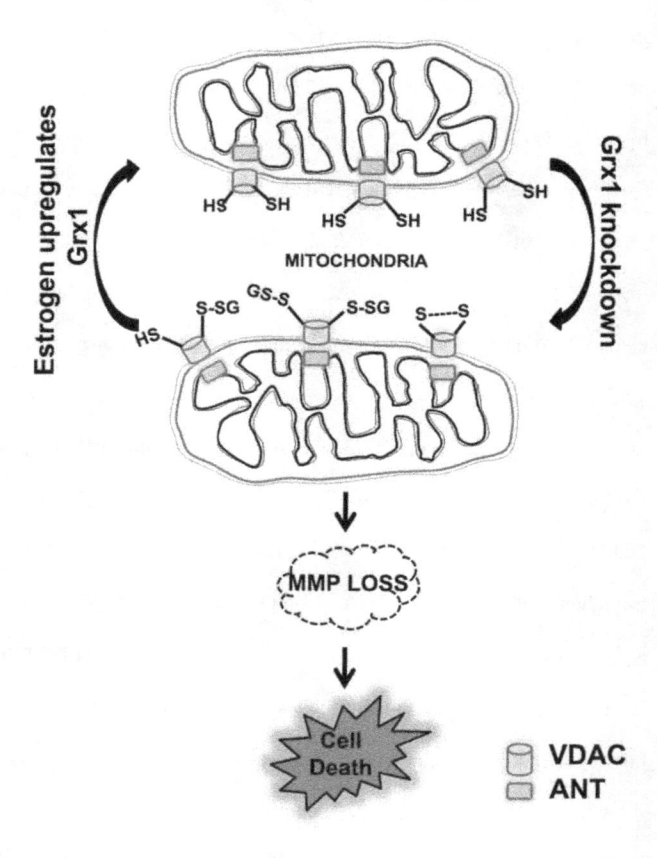

Figure D2: Cytosolic Grx1 downregulation results in loss of mitochondrial membrane potential.
Downregulation of cytosolic Grx1 leads to modification of critical thiol groups of the outer mitochondrial membrane protein VDAC, resulting in loss of membrane potential which could eventually lead to cell death. Grx1 expression is regulated by estrogen and its upregulation by estrogen prevents MMP loss by maintaining redox status of critical thiol groups in the mitochondria.

5) ***Loss of DJ-1 by knockdown of glutaredoxin (Grx1), triggers translocation of Daxx and ensuing cell death.***

DJ-1 is one of the putative genes linked to recessive familial form of PD. Loss of function mutations such as deletion, truncation and point mutations L166P, M26I, E64D etc. (Gorner et al., 2004) in the DJ-1 locus account for 1 to 2% of early onset PD. Studies using the loss of function mutations have suggested various functions for DJ-1, such as an antioxidant, a transcriptional co-activator and/or a molecular chaperone. It acts as an antioxidant by scavenging reactive oxygen species (ROS) and thereby helps maintain the redox status of other proteins (Mitsumoto et al., 2001; Ooe et al., 2005; Lev et al., 2008). DJ-1 has multiple cysteine residues whose oxidation enables it to act as a redox sensor during oxidative stimuli, such as hydrogen peroxide or paraquat (Mitsumoto et al., 2001; Taira et al., 2004).

Protein thiol homeostasis is critical for normal functioning of proteins and it may be perturbed during an intracellular oxidizing environment. Cells harbor different antioxidant systems to regulate redox environment and glutathione (GSH) is the most abundant cellular thiol antioxidant, which either scavenges ROS directly, or mediates glutathionylation, wherein GSH reversibly associates with the thiol groups of proteins to form mixed protein-glutathione disulfides 'PrSSG' (Mieyal et al., 2008). Hence, glutathionylation is an essential mode of regulation of redox status of proteins and thiol disulfide oxidoreductases (TDORs) are responsible for protecting and/or modulating protein function primarily by reducing thiol disulfides.

Insights from the previous study have indicated that glutaredoxin, a TDOR, is crucial for modulating the redox status of critical cysteine residues in proteins and its downregulation perturbs the thiol homeostasis leading to loss of mitochondrial membrane potential. With this rationale, in this project we studied the effect of perturbation of thiol homoeostasis on DJ-1. Thiol homeostasis was perturbed by downregulating Grx1, a key enzyme involved in glutathionylation and deglutathionylation in order to maintain redox homeostasis during adverse oxidative conditions. Strikingly, we observed that the levels of DJ-1, a key antioxidant protein were decreased by Grx1 knockdown. This was unusual considering that DJ-1 is a stress response protein which is upregulated under a variety of cellular stress models (Lev et al., 2008; Lev et al., 2009; Sakurai et al., 2009). Moreover, loss of DJ-1 was post-translational since its mRNA levels were unaffected. Glutathione (GSH) is the most abundant antioxidant present in cells and is readily utilized for maintaining thiol redox status during oxidative conditions (Wu et al., 2004). Its depletion by L-buthionine-S,R-sulfoximine (BSO) is known to generate reactive oxygen species and cause oxidative stress (Merad-Boudia et al., 1998; Gabryel and Malecki, 2006). We observed that GSH depletion using BSO caused enhanced generation of reactive oxygen species in our system and also resulted in loss of mitochondrial membrane potential, similar to what has been observed by Grx1 knockdown. However it did not cause cell death at the concentration we used (100 µM) as has been reported earlier also (Marengo et al., 2008a; Marengo et al., 2008b; Park et al., 2009b). Considering this fact, we used GSH depletion using BSO, as a parallel paradigm to induce oxidative stress and study its effects on DJ-1 translocation and downstream mechanisms.

DJ-1 is reported to translocate to mitochondria when exposed to an oxidative challenge such as rotenone, MPP$^+$ or H_2O_2 and this is considered to be protective mechanism, however further understanding is warranted in this regard (Canet-Aviles et al., 2004; Ashley et al., 2009; Junn et al., 2009). We looked at the sub-cellular localization of DJ-1 in response to both stressors used and strikingly found a discrepancy, wherein amid the two paradigms, Grx1 knockdown alone but not GSH depletion led to mitochondrial translocation of the residual DJ-1.

DJ-1 sequesters death associated protein, Daxx, in the nucleus in promyelocytic leukemia (PML) bodies preventing its nuclear export and subsequent activation of ASK1 mediated, and other cell death pathways in the cytosol (Junn et al., 2005). Consistent with this role of DJ-1, we found increased amount of Daxx in the cytosol following Grx1 knockdown, whereas it remained unaffected following GSH depletion. To further confirm whether nuclear export of Daxx was triggered by decrease in nuclear DJ-1 protein, we expressed WT DJ-1 ectopically. Even though, the levels of the overexpressed WT human DJ-1 were also depleted remarkably during Grx1 knockdown, the residual ectopic DJ-1 was sufficient to augment the endogenous levels and prevent nuclear export of Daxx, thereby affording protection. The role of Daxx in cell death and its action following translocation has been a matter of debate (Junn et al., 2005; Roubille et al., 2007; Zobalova et al., 2008; Waak et al., 2009). However, the results obtained in current study demonstrate that translocation of Daxx to the cytosol from the nucleus occurs only when nuclear DJ-1 levels decrease. Hence, cytosolic translocation of Daxx is dependent in turn on nuclear export of DJ-1. Daxx translocation to cytosol was also accompanied by cell death, again only in the paradigm wherein Grx1 was downregulated but not during GSH depletion. This is in concurrence with earlier reports using BSO (Park et al., 2009a). Ectopic expression of human WT DJ-1, besides preventing Daxx translocation, could also abolish cell death.

DJ-1 has three critical cysteine residues at amino acid position 46, 53 and 106. Since these residues are known to play an important role in the anti-oxidant function of DJ-1 (Mitsumoto et al., 2001; Canet-Aviles et al., 2004; Kinumi et al., 2004; Waak et al., 2009) and Grx1, being a TDOR could potentially act on these cysteine molecules, we examined the status of DJ-1 cysteine mutants (C53A and C106A) following Grx1 knockdown. Both the DJ-1 mutant proteins did not suffer any loss in response to Grx1 knockdown indicating that the cysteine residues played a critical role in the loss of DJ-1 and were being potentially modulated by Grx1. Oxidative modification of these residues that could potentially occur during Grx1 knockdown may adversely affect the stability of the protein resulting in its degradation. Thus, both the cysteine residues at 53 and 106 positions that were examined are critical for cellular protection under the current experimental paradigm.

DJ-1 could potentially be oxidatively modified under decreased Grx1 levels, thus leading to its degradation, a phenomenon not elicited by GSH depletion. To infer if DJ-1 was oxidatively modified in absence of Grx1, we examined the redox status of DJ-1 in response to Grx1 knockdown by derivatizing the free and reduced thiol groups with an alkylating agent and then studying the shift in migration of reduced and oxidized DJ-1 protein. We found that the intact DJ-1 in the cells was essentially present in the reduced form since there was a proportional decrease in the reduced fraction of DJ-1 with a loss in total protein levels. This suggests that following oxidative modification, DJ-1 is rapidly degraded and the residual DJ-1 present in cells is principally in the reduced form.

Under oxidative intracellular conditions, as a primary step, DJ-1 is presumably oxidized to sulfenic acid at the critical cysteines in both the paradigms used in this study (Miyazaki et al., 2008; Tsuboi et al., 2008). Sulfenic acids can then form mixed disulfide with GSH or other protein thiol(s), reactions catalyzed by thiol disulfide oxidoreductases including Trx and Grx. Grx1 would thus play critical roles in both glutathionylation and deglutathionylation of DJ-1. In the absence of Grx1, the oxidation of DJ-1 would proceed in a stepwise manner from the reversible sulfenic acid state (-SOH) to the irreversible higher oxidation states - sulfinic (-SO$_2$H) and sulfonic acid (-SO$_3$H). Once this occurs,

DJ-1 presumably undergoes translocation (Canet-Aviles et al., 2004; Blackinton et al., 2009) and degradation. Provided that GSH levels are depleted following BSO, the sulfenic acids are presumably transformed predominantly to protein mixed disulfides (PrSSPr) and also to PrSSG, which are eventually restored back to reduced thiols (PrSH) by Grx1, thereby preventing further irreversible oxidation of DJ-1 and thereby subsequent lethal events downstream. Thus, the presence of Grx1 is critical to prevent irreversible oxidation of thiol groups in DJ-1, while in conditions wherein GSH is depleted the protein thiol pool may be effective in buffering the sulfenic acids from further oxidation through mixed disulfide formation (Fig. D3).

Grx1 downregulation   GSH depletion

DJ-1 $\xrightarrow{ROS}$ DJ-1-SOH   ...(1)... DJ-1 $\xrightarrow{ROS}$ DJ-1-SOH

DJ-1-SOH $\xrightarrow{ROS}$ DJ-1-SO$_2$H ...(2)... DJ-1-SOH + Pr-SH + GSH (residual)

DJ-1-SS-Pr + DJ-1-SSG $\longleftarrow$ Grx/Trx/NE
Major product   Minor product

DJ-1-SO$_2$H $\xrightarrow{ROS}$ DJ-1-SO$_3$H ...(3)... DJ-1-SS-Pr + DJ-1-SSG

Grx | Trx
↓
DJ-1-SH

Figure D3: Schematic representation of oxidative modifications occurring in response to GSH depletion and Grx1 downregulation.

Both Grx1 knockdown and GSH depletion generate ROS. ROS generation causes the reversible oxidative modification of DJ-1-SH (reduced thiol) to DJ-1-SOH (sulphenic acid) in both paradigms (Eqn. 1). During Grx1 knockdown, thiol groups of a fraction of DJ-1 pool undergo irreversible sulphoxidation in a stepwise manner forming DJ-1-SO$_2$H (sulphinic acid; Eqn. 2) and DJ-1-SO$_3$H (sulphonic acid; Eqn. 3) due to absence of Grx1 and enzymatic glutathionylation thereafter. During GSH depletion, oxidative modification proceed to DJ-1-SOH (Eqn.1) but leads to formation of protein mixed disulphides (DJ-1-SSPr) and DJ-1-glutathione mixed disulphides

(DJ-1-SSG; Eqn. 2; catalyzed by Grx, Trx and/or non-enzymatic NE) which are consequently reversed to reduced DJ-1 thiols (DJ-1-SH) by Grx or Trx (Eqn. 3).

In order to further understand the mechanism underlying the loss of DJ-1, we studied the mechanism of its degradation by inhibiting the ubiquitin-proteasome machinery and examining the effect on both DJ-1 endogenous and ectopically expressed protein. Depletion of endogenous DJ-1 levels could not be prevented by inhibiting ubiquitin-proteasome machinery thus ruling out its proteasomal degradation, however loss of ectopically expressed FLAG tagged WT DJ-1, could be attenuated by proteasome inhibitors, indicating that the constitutive and overexpresssed proteins were processed by different mechanisms following their oxidative modification at critical cysteine residues. Whereas FLAG tagged WT DJ-1 was partially processed by ubiquitin proteasome machinery, degradation of endogenous DJ-1 was occurring by an unknown mechanism which is yet to be deciphered. In concurrence with this result, we found greater loss of ectopically expressed human WT DJ-1 following Grx1 knockdown in cells (~50%), which was to a great extent more than the observed loss of the endogenous protein (~25%). There are evidences of DJ-1 acting as a protease in a cell-free system (Honbou et al., 2003; Gorner et al., 2004; Olzmann et al., 2004) thus suggesting that DJ-1 upon oxidation can undergo auto-proteolysis. Such a possibility in this regard cannot be ruled out and needs to be probed further. We also could not find a direct interaction between Grx1 and DJ-1 under normal conditions, thereby suggesting that Grx1 either indirectly modulates the redox status of DJ-1 or forms a brief intermediate complex for the modulation of its redox status.

Thus, in the current study we witness that a small decrement of DJ-1 (~25-30%) triggers a cascade of deleterious responses indicating the importance of maintaining critical levels of this protein in the specific sub-cellular compartments (Fig. D4). Several important cellular functions such as, anti-oxidant (Ooe et al., 2005; Lev et al., 2008), transcriptional regulator (Clements et al., 2006; Davidson et al., 2008; Zhong and Xu, 2008), chaperone activity (Shendelman et al., 2004; Batelli et al., 2008; Liu et al., 2008a) have been attributed to DJ-1. Dysfunction or down-regulation of DJ-1 potentiates cellular toxicity (Bretaud et al., 2007) while over-expression is protective (Shinbo et al., 2006; Mo et al., 2008). Given that loss of function mutations in DJ-1 are associated with familial PD, the importance of this multi-functional protein cannot be over-emphasized. This study solely demonstrates that perturbation of protein thiol homeostasis through downregulation of a TDOR, Grx1 can lead to loss of DJ-1 perhaps via degradation. Downregulation of Grx1 adversely affects mitochondrial complex I activity and increases the vulnerability of SNpc cells in animal model of PD (Kenchappa et al., 2002; Kenchappa and Ravindranath, 2003; Diwakar et al., 2007). Conceivably, perturbation of protein thiol homeostasis caused by dysfunction of TDORs, including Grx1 may confer susceptibility to sporadic PD through multiple mechanisms including loss of DJ-1.

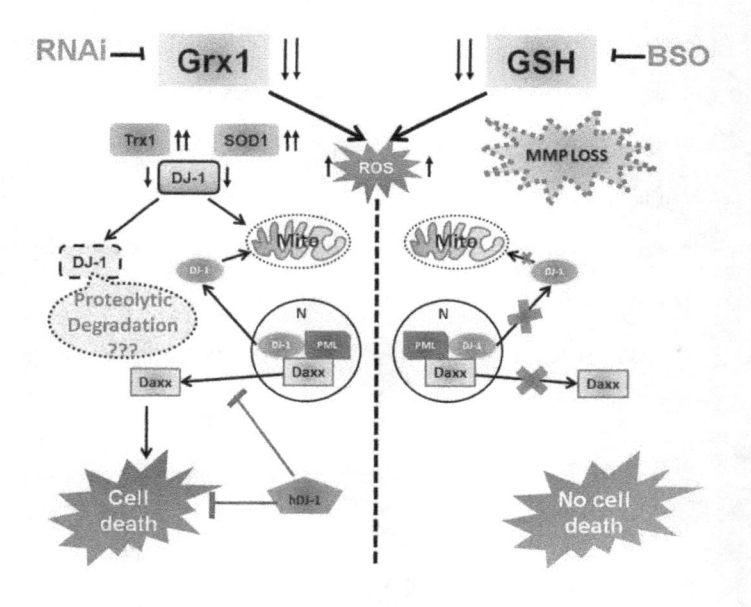

Figure D4: Schematic representation of differential phenomenon occurring in response to GSH depletion and Grx1 downregulation.

Both Grx1 knockdown and GSH depletion trigger generation of reactive oxygen species. Grx1 knockdown causes depletion of DJ-1 and only Grx1 knockdown leads to mitochondrial translocation of residual DJ-1, nuclear export of Daxx and consequent cytotoxicity. Ectopic WT DJ-1 (hDJ-1) prevents Daxx translocation and subsequent cell death. (N - nucleus; M - mitochondria).

6) *Mitochondrial Glutaredoxin (Grx2) confers protection against MPP$^+$ mediated cell toxicity.*

Mitochondrial dysfunction has been implicated in several neurodegenerative disorders including PD (Ohta and Ohsawa, 2006; Park et al., 2006). It results from oxidative modification of cellular macromolecules and reduced levels of antioxidants as has been detected in brains of PD patients (Jenner, 1998). Hence, maintenance of mitochondrial function during oxidative stress would require the presence of an efficient antioxidant system, which can reduce protein disulfides to active thiols, within the mitochondria.

Mitochondrial glutaredoxin 2 (Grx2), a mitochondrial counterpart of Grx1, is a glutathione-dependent oxidoreductase which deglutathionylates mitochondrial proteins such as complex I, that forms mixed disulfides during oxidative conditions (Beer et al., 2004). It is reported to afford protection against oxidative stress induced by mitochondria (Daily et al., 2001b, a) and its overexpression decreases susceptibility to apoptosis (Enoksson et al., 2005). Grx2 is presents in both, nucleus and mitochondria, and has an active site consisting of Cys-Ser-Tyr-Cys that helps maintain redox homeostasis within mitochondria. The four cysteine residues present in two molecules of Grx2 are coordinated with iron in a non oxidizable $[2Fe-2S]^{2+}$ cluster to form an inactive dimeric holo Grx2 (Lillig et al., 2005). The cluster is preserved as such and remains inactive in the presence of glutathione, while in presence of oxidants such as glutathione disulfide the dimer dissociates, thereby leading to the formation of the monomeric active Grx2 enzyme.

The iron–sulfur cluster perhaps serves as a redox sensor and activates Grx2 during oxidative stress.

MPP^+ is the active component of MPTP, a model neurotoxin which generates ROS, results in oxidative stress thereby, inhibiting mitochondrial complex I. We used MPP^+ to study if Grx2 could potentially prevent Neuro-2a cells from MPP^+ induced cytotoxicity. Grx2 was overexpressed in Neuro-2a cells, which were then treated with MPP^+. Overexpression of Grx2 completely ameliorated the apoptotic cell death mediated by MPP^+ as has been reported earlier, indicating the potential neuroprotective effects of Grx2. Thus, besides Grx1, Grx2 is also important for the maintenance of mitochondrial function and its integrity. Single nucleotide polymorphisms which adversely alter the activity of Grx2 may increase susceptibility to neurodegenerative disorders wherein oxidative stress is a key player.

II. **Mechanisms of cell death in Parkinson's disease and role of estrogen.**

2) *MPP^+ induced activation of p38 in primary culture of human dopaminergic neurons.*

Previous studies from our lab have demonstrated that MPTP induced complex I inhibition could be completely prevented by administration of α-lipoic acid to mice prior to injecting MPTP but death of SNpc dopaminergic neurons was only attenuated. It was also demonstrated that besides inhibition of complex I, MPP^+, was also inducing MAP kinase pathways, as shown by the activation of ASK1, which led to the demise of SNpc dopaminergic neurons (Karunakaran et al., 2007b). MPTP induces activation of p53 and Bax, which play a role in selective cell death of dopaminergic cells in SNpc and inhibiting p53 and Bax by genetic ablation or RNA interference prevented the ensuing cell death (Duan et al., 2002; Eberhardt and Schulz, 2003; Fei and Ethell, 2008). Both p53 and Bax knockout mice showed lowered neuronal loss following MPTP treatment as compared to their wild-type counterpart and Bax knockouts were more protective amid the two. There were, however no reports clearly demarcating the events upstream to p53 activation. Several *in vitro* studies performed in cultured cells demonstrated the phosphorylation of JNK and the neuroprotection offered by JNK inhibitors (Wang et al., 2004a; Wang et al., 2004b), however, failure of clinical trials using the JNK inhibitor CEP1347 question its efficacy and the role of JNK in the demise of dopaminergic cells in SNpc. Further, microarray studies done earlier in our laboratory revealed sustained upregulation of p38 following acute and subchronic doses of MPTP. This gave us an insight to probe further into the cascade downstream to ASK1 activation which leads to activation of p53 mediated apoptosis following MPP^+ treatment. In this study, we used cultured dopaminergic neurons derived from human neural progenitor cells as our model system, to discern the role of p38 in MPP^+ mediated cytotoxicity. The culture was found appropriate for use since it had approximately 90% of neuronal population of which 97% stained positive for tyrosine hydroxylase, indicating their dopaminergic lineage. Furthermore, it also stained positive for several molecules involved in apoptotic signalling such as ASK1, p38, JNK, p53 and also for dopamine transporters (DAT). Both MPTP and MPP^+ were used as model neurotoxins initially but MPP^+ (10 µM) worked better since the culture population was predominantly neuronal. We carefully studied the MPP^+ mediated cell death cascade by scrutinizing the signalling at each step using specific inhibitors for p38, JNK and p53 and examined the effect of MPP^+ toxicity in cultured dopaminergic neurons. Treatment of MPP^+ to the cells triggered apoptotic cascade causing activation of p38, phosphorylation and translocation of p53 to the nucleus and subsequent cytotoxicity as measured by TUNEL assay. p38 is known to phosphorylate p53 downstream and this post-translational modification facilitates p53 accumulation in the nucleus (Shieh et al., 1997; Vitale et al., 2008). We observed that p38 inhibitor; SB239063, which inhibits both p38 α and β, prevented the translocation of p53 to the nucleus and afforded complete protection against cell death in these cells. Further, JNK inhibitor, SP600125 and pifithrin (p53 inhibitor), also conferred protection against MPP$^+$ mediated cell death, however, SP600125 could not completely prevent nuclear translocation of p53. The antioxidant, α-lipoic acid could only attenuate the cytotoxicity caused by MPP$^+$, thereby confirming the activation of apoptotic signaling in the dopaminergic cultured cells. Thus, these cells derived from human progenitors, offered an excellent *in vitro* model for discerning molecular mechanisms after a neurotoxic insult. Besides this, we also studied the effect of MPP$^+$ toxicity in human neuroblastoma cell line (SH-SY5Y) and observed that, MPP$^+$ induced translocation of Daxx from the nucleus to the cytoplasm could be prevented by pre-treatment of cells with antioxidants such as α-lipoic acid and N-acetyl cysteine, further confirming the propagation of several cell death mechanism induced by MPP$^+$ neurotoxicity.

Identification of signaling cascades occurring in the affected subset of cell is vital for discovering and developing therapeutic strategies that can offer neuroprotection and are promising enough for modifying the disease, by slowing down the degeneration of vulnerable cell population. It is quite apparent from a global perspective that several mechanisms contribute to the neurodegeneration seen in diseases like PD and hence JNK inhibitor, CEP1347 alone, failed to slow down the progression of the disease in clinical trials. The progression of death signaling cascade as seen in our model indicates that activation of p38 is also potentially an important contributor to the disease process leading to the demise of cultured dopaminergic neurons, however, efficacy of specific p38 inhibitors in slowing down the progression of disease is yet to be determined. In the present scenario, a combinatorial approach targeting molecules in several pathways simultaneously operating in different cell populations would boost neuroprotection.

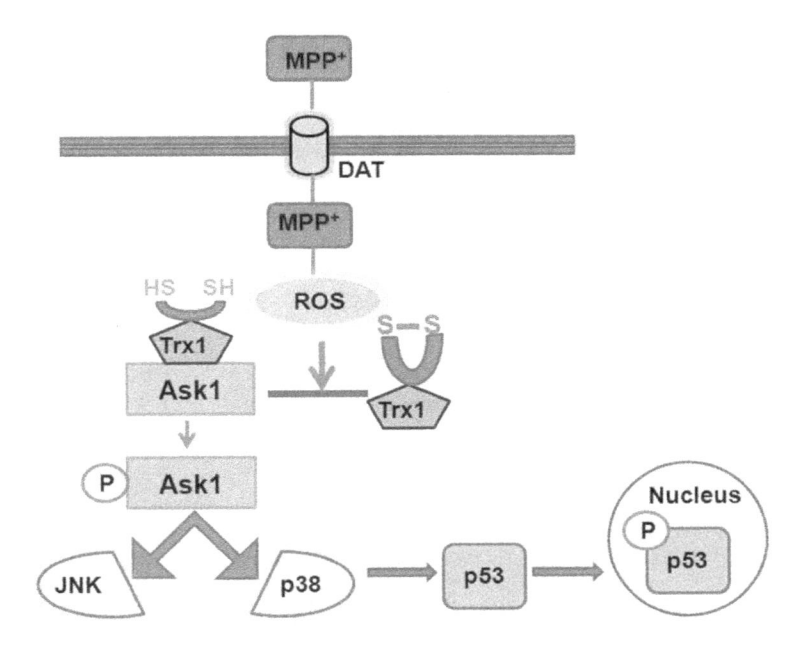

Figure D5: Schematic representation of signaling cascade induced by MPP$^+$ toxicity.

MPP+ after entering the neurons through dopamine transporters (DAT) induces ROS which triggers redox dependent apoptotic signaling. Trx, which normally in its reduced form remains associated with ASK1 gets oxidized thus causing phosphorylation of ASK1 and rendering it free to propagate the downstream signaling. Activation of ASK1 in turn activates JNK and p38, downstream to it, and p38 activation further leads to phosphorylation of p53 and its translocation to nucleus where it regulates the expression of genes which lead to cell death.

3) *Redox driven cell death signaling cascade activated in male mice are attenuated in females mice following MPTP exposure*

Incidence and prevalence of PD has been reported to be more in men as compared to women with an overall incidence ratio of 1.46 in males vs. females (Shulman, 2007). The gender difference seen in human population is also reflected in the animal models wherein female mice are protected from MPTP toxicity. Previous studies from our lab have demonstrated that whereas, male mice were vulnerable to MPTP induced toxicity, females were unaffected (Kenchappa et al., 2004), however, the underlying mechanisms which confer neuroprotection to females were quite unclear. In the current study, we explored the differential molecular mechanism occurring in male and female mice after MPTP administration. We looked at the mechanisms underlying neuroprotection seen in females by examining the redox activated MAP kinase signaling cascades that have been implicated in the selective degeneration of SNpc dopaminergic neurons in mice (Karunakaran et al., 2007b; Karunakaran et al., 2008).

We hereby demonstrate that the degeneration of dopaminergic neurons in SNpc following MPTP is attenuated significantly in female mice as compared to males. Earlier studies on sex difference in MPTP toxicity have shown that striatal dopamine levels as well as the SNpc dopaminergic neurons and their terminals are protected in female mice, which is abolished by ovariectomy (Freyaldenhoven et al., 1996; Liu et al., 2008b; Morissette et al., 2008). Thus, the sex difference in response to MPTP toxicity is not restricted to striatum alone but also extends to the preservation of the dopaminergic neurons in SNpc as shown in the present study. Contribution of altered metabolism of MPTP or uptake of MPP^+ is unlikely in the sex difference observed in this study, since C57 black female mice are known to metabolize MPTP in a manner similar to males. Further, since DAT (dopamine transporter) and VMAT2 (vesicular monoamine transporter) expression is higher, the uptake of MPP^+ may in fact be greater in females (Dluzen and McDermott, 2008).

Another interesting observation is the fact that mitochondrial dysfunction seen as complex I inhibition is completely absent in females but some degree of cell loss is seen after 8 days of MPTP (Fig. 43B), thereby indicating that in addition to the mitochondrial dysfunction, a parallel pathway of cell loss is operative in MPTP treated animals. This is similar to the observation made in male mice pretreated with α-lipoic acid (ALA) prior to MPTP (Karunakaran et al., 2007b), wherein, the mitochondrial dysfunction was abolished by ALA but the cell loss was only attenuated. Earlier studies from our laboratory have shown that redox modification of critical thiols can lead to aberrant signaling and initiation of the death signaling cascade by the MAPKKK, ASK1 (Karunakaran et al., 2007b). We therefore, examined if such a cascade was operative in females after MPTP treatment.

In order to inspect the early events after MPTP treatment, we chose to study the MAPK signaling cascade at early time points after a single dose of MPTP. We found that ASK1 was not activated in ventral midbrain of female mice unlike males (Fig. 44A). Since the levels of reduced thioredoxin are critical for suppressing ASK1 activation, we examined the constitutive expression of Trx in female midbrain and found that Trx, TR and GR were constitutively expressed in higher amounts in females similar to the fact stated earlier that estrogen regulates the expression of enzymes having antioxidant properties. Trx is oxidized during oxidative stress and TR helps in reducing oxidized Trx thus maintaining the homeostasis. Thus, higher levels of Trx in female mice could potentially prevent the autophosphorylation of ASK1 and the downstream activation of MAP kinases, such as p38. Other than this, we also observed the absence of p38 activation in the TH positive neurons in SNpc of female mice, quite unlike to that seen in males. Earlier studies from our laboratory have documented that p38 is activated selectively in the dopaminergic neurons of SNpc following MPTP and this leads to phosphorylation and transactivation of p53 and enhanced expression of pro-apoptotic genes, such as Bax and Puma (Karunakaran et al., 2008) downstream. We did not find activation of p38 and propagation of downstream cascade in SNpc of female mice potentially due to the high levels of Trx.

ASK1 can also propagates the death cascade by interacting with Daxx, a transcriptional repressor that is normally present in the nucleus (Song and Lee, 2003a). Daxx, a death associated protein remains associated with DJ-1, a redox sensing protein within the PML (promyelocytic leukaemia) bodies in the nucleus. DJ-1 sequesters Daxx within the nucleus thereby preventing the nuclear export of Daxx (Junn et al., 2005). Following MPTP administration there is loss of nuclear DJ-1 which results in its release from the nucleus, translocation to the cytosol and its subsequent association with ASK1 (Karunakaran et al., 2007b). Intriguingly, nuclear DJ-1 levels in female mice are unaffected after MPTP administration and Daxx translocation to the cytosol does not occur at all. Further, it was a fascinating observation, that the constitutive levels of DJ-1 were higher in female mice when compared to males. This is significant considering the fact that mutations in DJ-1 have been implicated in familial PD cases (Djarmati et al., 2004).

Even though the incidence of PD is known to be lower in women and significant neuroprotection is seen in female rodents in animal models, the underlying molecular mechanisms were unclear. It is unlikely that estrogen has a direct anti-oxidant effect since the concentrations required for such action are several fold higher than the physiological levels (Santanam et al., 1998). It has been proposed that estrogen mediates its action through its receptors, α and β by a variety of mechanisms including transcription of critical genes which are neuroprotective, such as BDNF, IGF1 and Bcl2 (Nilsen and Diaz Brinton, 2003). Significance of this study lies in the fact that we demonstrate here for the first time, that constitutive expression of Trx, TR and GR is higher in the brain of female mice, which in turn prevents aberrant signaling mediated through ASK1. Failure of ASK1 activation further prevents the downstream cascades mediated through p38 phosphorylation. Further, we also demonstrate that DJ-1 is expressed at higher levels constitutively in the midbrain of female mice, which prevents the translocation of Daxx and its interaction with ASK1 in the cytosol (Fig. D6). This study would potentially help us to understand the mechanisms involved in the neuroprotection seen in females and to develop theraupetic strategies towards maintaining redox homeostasis. Hence, small molecules targeted at increasing the expression of critical redox regulating proteins such as Trx, TR and DJ-1 may help protect the SNpc dopaminergic neurons and slow down the progression of the disease.

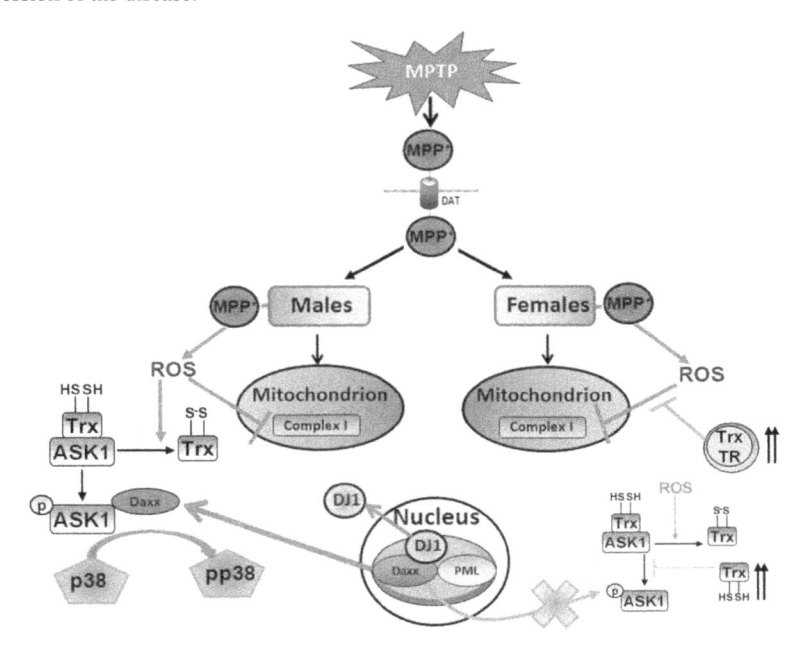

Figure D6: Schematic representation of differential activation of apoptotic pathways in male and female mice in response to MPTP.
1-methyl-4-phenylpyridinium (MPP$^+$), the toxic metabolite of 1-methyl-4-phenyl-1, 2, 3, 6 tetrahydropyridine (MPTP) causes increased production of reactive oxygen species (ROS) and mitochondrial dysfunction in dopaminergic neurons by inhibiting complex I of the electron transport chain. In male mice treated with MPTP, ROS activates apoptosis signal regulating kinase (ASK1) through the oxidation of thioredoxin (Trx), which results in its dissociation from ASK1. As a consequence, the downstream kinase p38 is phosphorylated. ROS also triggers decrease in levels of nuclear DJ-1 and subsequent translocation of Daxx from the nucleus to the cytosol. Interaction of Daxx with ASK1 in the cytosol further propagates the death cascade. Activation of MAP kinase (ASK1 & p38) cascade and translocation of Daxx is suppressed in females due to enhanced thioredoxin system and retention of DJ-1 and sequestration of Daxx within the nucleus.

Hence, we observed the significance of redox regulation in the normal physiology and functioning of cell. We also monitored, how the dysregulation of redox homeostasis could trigger redox driven apoptotic cascades thus leading to the demise of dopaminergic neurons in SNpc, in ventral midbrain of male mice administered with MPTP. Further we also comprehend the role of thiol disulfide oxidoreductases, especially the enzymes involved in glutaredoxin and thioredoxin system in the maintenance of redox homeostasis and henceforth in conferring the neuroprotection observed in female mice by preventing the activation of the redox dependent cell death cascades following MPTP administration. Furthermore, we also came across the fact that how a diminutive decline in the levels of glutaredoxin 1, a cytosolic protein, could potentially affect the cellular physiology, thus leading to the loss of mitochondrial membrane potential and thereby causing mitochondrial dysfunction. This study, for the first time elucidates to an extent, the role of a cytosolic protein (Grx1) in the maintenance of mitochondrial complex I as observed earlier in the studies from our lab (Kenchappa et al., 2004; Diwakar et al., 2007). What was even more striking was the finding that knockdown of Grx1 causes oxidative modification of a redox sensitive protein DJ-1, implicated in familial PD, thereby causing its loss of function. It gives us an insight to consider that the oxidative conditions occurring in neurodegenerative diseases, specifically PD, may be responsible for causing irreparable and deleterious modifications in certain redox sensitive proteins, thereby leading to their loss of function and thus augmenting the sporadic form of pathology. Finally, the work done towards this thesis is a humble attempt to unveil the significance of redox regulation and role of thiol disulfide oxidoreductases in maintaining the redox homeostasis which is indispensible for normal physiology and survival of the cell.

SUMMARY AND CONCLUSIONS

The main objective of the study described in this thesis was to examine redox regulation by thiol disulfide oxidoreductases and implications of redox driven cell death mechanisms in PD. Thus the study was carried out to address the following specific aims:

1. To investigate the mechanistic role of cytosolic glutaredoxin in maintaining mitochondrial function.
2. To examine if oxidative stress generated by Grx1 knockdown has an effect on redox sensitive protein DJ-1 (implicated in familial PD) and if Grx1 is involved in the modulation of its redox status.
3. To examine if mitochondrial glutaredoxin offers protection against MPP^+.
4. To analyze apoptotic signaling triggered in response to MPP^+, the toxic metabolite of MPTP, using primary dopaminergic neuronal cultures derived from human CNS progenitor cells as the model system.
5. To understand the mechanism underlying estrogen mediated neuroprotection using MPTP mouse model for PD.

The conclusions derived from the studies are as follows:

1) Knockdown of cytosolic glutaredoxin 1 leads to loss of mitochondrial membrane potential: Implication in neurodegenerative diseases.

Grx1, a cytosolic enzyme was known to influence the mitochondrial function but however, its mode of action was not very clear. To explore its role in maintaining the mitochondrial function, an *in vitro* approach was adopted and its role was studied by overexpressing and silencing it in neuroblastoma cell lines (SHSY-5Y and Neuro-2a) of human and mouse origin respectively.

Grx1 is localized in cytoplasm as opposed to Grx2, which is present in mitochondria. Silencing of Grx1 using shRNA resulted in consistent 50-60% loss of Grx1 at mRNA and protein level. It also generated reactive oxygen species and resulted in loss of mitochondrial membrane potential (MMP) in both SHSY-5Y and Neuro-2a cells as studied both qualitatively and quantitatively. This loss of MMP was successfully prevented by pretreating the cells with cyclosporine A, a permeability transition pore blocker and α-lipoic acid, a potent antioxidant. L-BOAA, an excitatory amino acid, which also causes mitochondrial complex I dysfunction was used as a parallel paradigm to induce stress and to understand the role of redox regulation in the maintenance of MMP. Treatment of SHSY-5Y cells with L-BOAA resulted in a similar loss of MMP and this loss could be prevented by overexpressing Grx1. SHSY-5Y cells were used to study the regulation of Grx1 by estrogen, since they express estrogen receptors α and β. Cells were treated with 17-β estradiol prior to the treatment with L-BOAA and measurement of MMP. Treatment of cells with estrogen resulted in upregulation of Grx1 and also prevented loss of MMP observed by L-BOAA treatment. Enhancing Grx1 protein level in cells, either by treatment with 17-β estradiol or Grx1 overexpression afforded protection against MMP loss caused by L-BOAA toxicity and Grx1 knockdown.

Glutaredoxin is known to modulate the redox status of proteins by glutathionylating the critical thiol groups during oxidative stress and then deglutathionulating them, once the stress subsides, thereby protecting them from irreversible oxidative modification. To examine the mechanism of maintenance of MMP and thereby mitochondrial function by Grx1, redox status of voltage dependent anion channel (VDAC) was studied in response to Grx1 knockdown. Grx1 knockdown resulted in decrease in fraction of VDAC, having reduced thiol groups, thus indicating that depletion of Grx1 protein levels resulted in oxidative modification of VDAC thus resulting in the opening of permeability transition pore and loss of MMP and henceforth the mitochondrial function (Saeed et al., 2008). However, adenine nucleotide translocase (ANT), an inner mitochondrial membrane protein, also a component of permeability transition pore was not adversely affected by Grx1 knockdown, thus indicating the importance of cytosolic glutaredoxin in maintaining the redox status of proteins present on outer mitochondrial membrane and thus the mitochondrial function.

2) *Loss of DJ-1 by knockdown of glutaredoxin (Grx1), triggers translocation of Daxx and ensuing cell death.*

DJ-1 is a redox sensitive protein and its loss of function mutations are implicated in pathogenesis of familial PD. To investigate if its redox modulation by increased oxidative environment impairs its function, oxidative conditions were created by downregulating Grx1 and effects of intracellular oxidative environment were studied on DJ-1, in Neuro-2a cells.

Grx1 knockdown caused enhanced ROS generation, resulting in oxidative stress and upregulation of stress response genes such as Trx and SOD1. Besides this, strikingly, it also caused depletion of DJ-1 at protein level but not at mRNA levels, suggesting that DJ-1 protein was sensitive to oxidative conditions created by Grx1 knockdown. Grx1 knockdown also triggered the translocation of residual DJ-1 to the mitochondria and induced nuclear export of Daxx and subsequent apoptotic cell death, a phenomenon not triggered by GSH depletion, an alternative paradigm used for global oxidative stress, which also generated ROS and resulted in loss of mitochondrial membrane potential like Grx1 knockdown. Overexpression of wild type (WT) DJ-1 completely prevented nuclear export of Daxx and also abolished the ensuing cytotoxicity, although, decrease in the proteins levels of ectopic WT DJ-1 was more compared to endogenous DJ-1. However levels of overexpressed cysteine modified and PD related mutants of DJ-1 (C106A, C53A and L166P) were unaffected suggesting that Grx1 knockdown modulates thiol groups of cysteine residue(s) and leads to its consequent depletion. Examination of redox status of DJ-1 in response to Grx1 knockdown revealed that loss of reduced DJ-1 was proportionate to loss in total protein, thereby predicting that oxidative modification of DJ-1 was resulting in its immediate clearance and the residual DJ-1 was present in its reduced form. On further probing the mechanism of degradation, it was found that only ectopic WT DJ-1 was being partially processed by ubiquitin proteasome machinery and the endogenous DJ-1 was being degraded by an unknown mechanism yet to be explored.

This study led to a new finding that the redox sensor, DJ-1 may be oxidatively modified through downregulation of Grx1, thereby resulting in its loss or impaired function which could be implicated in pathogenesis of sporadic PD.

3) *Glutaredoxin 2, the mitochondrial glutaredoxin offers protection against MPP$^+$ mediated cytotoxicity.*

Similar to cytosolic glutaredoxin (Grx1), mitochondrial glutaredoxin (Grx2) helps in maintaining the favourable redox environment, however it works within the mitochondria. MPP$^+$, the active metabolite of MPTP causes mitochondrial oxidative stress and results in inhibition of complex I in mitochondria thus causing subsequent cytotoxicity. Overexpression of Grx2 in Neuro-2a cells prevented MPP$^+$ induced cell death thus indicating the protective function of Grx2 within the mitochondria.

4) *MPP$^+$ induced activation of p38 in primary culture of human dopaminergic neurons.*

Primary cultures of human CNS progenitor cells were differentiated into neurons and characterised for the dopaminergic neuronal lineage by studying the expression of tyrosine hydroxylase (TH) and neuronal marker β-III-tubulin (Tuj). Approximately, 89.9% cells were found to be positive for both TH and Tuj as observed by immunostaining. These cells were used as a model system to decifer MPP⁺ induced cell death pathway using inhibitors for p38 (SB239063), JNK (SP600125) and p53 (pifithrin-α) and monitoring the nuclear translocation of p53 and consequent cell death.

SB239063, SP600125 and pifithrin-α conferred protection to these cells against MPP⁺ induced cytotoxicity. Probing into the mechanism of MPP⁺ induced cell death revealed the nuclear translocation of p53 which could be prevented by p38 inhibitor SB239063. JNK inhibitor, SP600125, however prevented the cell death but could only attenuate nuclear translocation of p53. (Karunakaran et al., 2008) Pretreatment of cells to α-lipoic acid also attenuated toxicity induced by MPP⁺ in these cells. Hence primary dopaminergic cultures derived from human CNS progenitor cells were an excellent model to decipher cell death pathways induced by MPP⁺.

5) *Redox driven cell death signaling cascade is alternatively activated in males and females.*

Incidence of PD has been reported to be lower in women when compared to men. Neuroprotective effects of estrogen although well recognized for decades, yet are not fully understood. To understand the mechanism underlying neuroprotection offered by estrogen, comparative evaluation of events was carried out in male and female mice, after MPTP administration. While MPTP induced inhibition of mitochondrial complex I was completely averted, the progressive degeneration of dopaminergic neurons occurring in SNpc of male mice was only attenuated in females. Besides this, MPTP failed to induce the activation of apoptotic signal regulating kinase 1 (ASK1), a MAPKKK in female mice whereas, ASK1 was significantly activated in their male counterparts, although the levels of total ASK1 were not different among the genders. On studying the expression of redox regulating enzymes, it was found that the constitutive expression of thioredoxin (Trx), thioredoxin reductase (TR) and glutathione reductase (GR) was higher in female mice compared to males.

On further exploring the signaling mechanism downstream to ASK1, p38, a MAP kinase, was found significantly activated in male mice whereas the response was completely suppressed in females, potentially due to the enhanced expression of Trx which prevents the autophosphorylation of ASK1 and its activation. Moreover Daxx, propagated the cell death signaling by its nuclear export in the SNpc of male mice administered with MPTP, a phenomenon lacking in females. Thus, in order to understand the neuroprotection observed in females, redox regulated cell death mechanisms were explored in male and female mice in response to MPTP. This study would potentially help in understanding the mechanisms involved in the neuroprotection seen in females and to develop strategies towards maintaining redox homeostasis which could potentially help in slowing down the progressive neurodegeneration seen in PD.

Work in this thesis, revolves around understanding the role of thiol disulfide oxidoreductases, particularly glutaredoxin, in maintaining redox homeostasis, importance of redox regulation and mechanism of propagation of cell death in PD due to impaired redox regulation. This work offers a key insight into the regulation of enzymes involved in redox maintenance by estrogen and molecular mechanisms of neuroprotection observed in female mice in animal models of Parkinson's disease.